Anonymous

Steam-Heating Problems

Anonymous

Steam-Heating Problems

ISBN/EAN: 9783743392670

Manufactured in Europe, USA, Canada, Australia, Japa

Cover: Foto ©berggeist007 / pixelio.de

Manufactured and distributed by brebook publishing software (www.brebook.com)

Anonymous

Steam-Heating Problems

Steam-Heating Problems;

OR,

QUESTIONS, ANSWERS, AND DESCRIPTIONS

RELATING TO

Steam-Heating and Steam-Fitting,

FROM

THE SANITARY ENGINEER.

WITH ONE HUNDRED AND NINE ILLUSTRATIONS.

NEW YORK:
THE SANITARY ENGINEER.

1886.

Copyright, 1886,
By The Sanitary Engineer.

THE SANITARY ENGINEER PRESS,
140 WILLIAM STREET, N. Y.

PREFACE.

THE SANITARY ENGINEER, while devoted to Engineering, Architecture, Construction, and Sanitation, has always made a special feature of its departments of Steam and Hot-Water Heating, in which a great variety of questions has been answered and descriptions of the work in various buildings have been given. The favor with which a recent publication from this office, entitled "Plumbing and House-Drainage Problems," has been received suggested the publication of "STEAM-HEATING PROBLEMS," which, though dealing with another branch of industry, is similar in character. It consists of a selection from the pages of THE SANITARY ENGINEER of questions and answers, besides comments on various problems met with in the designing and construction of steam-heating apparatus, and descriptions of steam-heating work in notable buildings.

It is hoped that this book will prove useful to those who design, construct, and have the charge of steam-heating apparatus.

TABLE OF CONTENTS.

BOILERS.

	PAGE
On blowing off and filling boilers,	17
Where a test-gauge should be applied to a boiler,	18
Domes on boilers: whether they are necessary or not,	19
Expansion of water in boilers,	21
Cast *vs.* wrought iron for nozzles and magazines of house-heating boilers,	21
Pipe-connections to boilers,	22
Passing boiler-pipes through walls: how to prevent breakage by settlement,	24
Suffocation of workmen in boilers,	25
Heating-boilers. (A problem.)	26
A detachable boiler-lug,	28
Isolating-valves for steam-main of boilers,	29
On the effect of oil in boilers,	33
Iron rivets and steel boiler-plates,	35
Proportions for rivets for boiler-plates,	37
Is there any danger in using water continuously in boilers?	38
Accident with connected boilers,	39
A supposed case of charring wood by steam-pipes,	40
Domestic boilers warmed by steam,	42

VALUE OF HEATING-SURFACES.

Computing the amount of radiator-surface for warming buildings by hot water,	45
Calculating the radiating-surface for heating buildings—the saving of double-glazed windows,	45
Amount of heating-surface required in hot-water apparatus boilers and in steam-apparatus boilers,	51

	PAGE
Calculating the amount of radiating-surface for a given room,	51
How much heating-surface will a steam-pipe of given size supply?	52
Coils vs. radiators and size of boiler to heat a given building,	53
Calculating the amount of heating-surface,	53
Computing the cost of steam for warming,	54

Radiators and Heaters.

A woman's method of regulating a radiator (covering it with a cosey),	56
Improper position of radiator-valves,	57
Hot-water radiator for private houses,	58
Remedying air-binding of box-coils,	59
How to use a stove as a hot-water heater,	60
"Plane" vs. "Plain" as a term as applied to outside surface of radiators,	61
Relative value of pipe on cast-iron heating surface,	62
Relative value of pipe on steam-coils,	62
Warming churches (plan of placing a coil in each pew),	65
Warming churches,	67

Piping and Fitting.

Steam-heating work—good and indifferent,	70
Piping adjacent buildings: Pumps vs. steam-traps,	71
True diameters and weights of standard pipes,	74
Expansion of pipes of various metals,	75
Expansion of steam-pipes,	75
Advantages claimed for overhead piping,	76
Position of valves on steam-riser connection,	77
Cause of noise in steam-pipes,	78
One pipe system of steam-heating,	79
How to heat several adjacent buildings with a single apparatus,	81
Patents on Mills' system of steam-heating,	83
Air-binding in return steam-pipes,	83
Air-binding in return steam-pipes, and methods to overcome it,	85

VENTILATION.

	PAGE
Sizes of registers to heat certain rooms,	86
Determining the size of hot-air flues,	88
Window ventilation,	90
Window ventilation,	91
Removing vapor from dye-house,	92
Ventilation of Cunard steamer "Umbria,"	93
Calculating sizes of flues and registers,	97
On methods of removing air from between ceiling and roof of a church,	98

STEAM.

Economy of using exhaust steam for heating,	99
Heat of steam for different conditions,	100
Superheating steam by the use of coils,	101
Effect of using a small pipe for exhaust steam-heating,	102
Explosion of a steam-table,	103

CUTTING NIPPLES AND BENDING PIPES.

Cutting large nipples—large in diameter and short in length,	106
Cutting crooked threads,	108
Cutting a close nipple out of a coupling after a thread is cut,	110
Bending pipe,	111
Cutting large nipples,	113
Cutting various sizes of thread with a solid die,	113

RAISING WATER AUTOMATICALLY.

Contrivance for raising water in high buildings,	116
Criticism of the foregoing and description of another device for a similar purpose,	118

Moisture on Walls, Etc.

	PAGE
Cause and prevention of moisture on walls,	120
Effect of moisture on sensible temperature,	122

Miscellaneous.

Heating water in large tanks,	124
Heating water for large institutions and high city buildings,	125
Questions relating to water-tanks,	127
Faulty elevator-pump connections,	128
On heating several buildings from one source,	130
Coal-tar coating from water-pipe,	130
Filters for feeding of house-boilers. Other means of clarifying water,	131
Testing gas-pipes for leaks and making pipe-joints,	132
Will boiling drinking-water purify it?	134
Differential rams for testing fittings and valves,	134
Percentage of ashes in coal,	136
Automatic pump-governor,	137
Cast-iron safe for steam-radiators,	138
Methods of graduating radiator service according to the weather,	139
Preventing fall of spray from steam-exhaust pipes,	144
Exhaust-condenser for preventing fall of spray from steam-exhaust pipes,	145
Steam-heating apparatus and plenum (ventilation) system in Kalamazoo Insane Asylum,	146
Heating and ventilation of a prison,	148
Amount of heat due to condensation of water,	152
Expansion-joints,	153
Resetting of house-heating boilers—a possible saving of fuel,	153
How to find the water-line of boilers and position of try-cocks,	155
Low-pressure hot-water system for heating buildings in England (comments by the *Sanitary Engineer*),	156
Steam-heating apparatus in Manhattan Company's and Merchants'; Bank Building, New York,	161

	PAGE
Boilers in Manhattan Company's and Merchants' Bank Building, with extracts from specifications,	172
Steam-heating apparatus in Mutual Life Insurance Building on Broadway,	177
The setting of boilers in Tribune Building, New York,	182
Warming and ventilation of West Presbyterian Church, New York City,	187
Principles of heating-apparatus, Fine Arts Exhibition Building, Copenhagen,	192
Warming and ventilation of Opera House at Ogdensburg, N. Y.,	195
Systems of heating houses in Germany and Austria,	198
Steam-pipes under New York streets—difference between two systems adopted,	203
Some details of steam and ventilating apparatus used on the continent of Europe,	206

MISCELLANEOUS QUESTIONS.

Applying traps to gravity steam-apparatus,	212
Expansion of brass and iron pipe,	213
Connecting steam and return risers at their tops,	213
Power used in running hydraulic elevators,	215
On melting snow in the streets by steam,	216
Action of ashes street fillings on iron pipes,	217
Arrangement of steam-coils for heating oil-stills,	217
Converting a steam-apparatus into a hot-water apparatus and back again,	219
Condensation per foot of steam-main when laid underground,	221
Oil in boilers from exhaust steam, and methods of prevention,	222

LIST OF ILLUSTRATIONS.

	PAGE
FIGURE 1.—Pipe connections to boilers..................................	23
" 2.—Passing boiler-pipes through walls to prevent breakage by settlement..	24
" 3.—Feeding boilers (a problem)...................................	26
" 4.—A detachable boiler-lug.......................................	29
" 5, 6, AND 7.—Isolating-valves..............................	30, 31
" 8.—The effect of oil in boilers...................................	34
" 9, 10, AND 11.—Proportions for rivets..........................	38
" 12.—Accident with connected boilers............................	39
" 13.—A supposed case of charring wood by steam-pipes...........	40
" 14.—Hot-water boilers for apartment-houses.......................	43
" 15.—A woman's method of regulating a radiator...................	56
" 16.—Improper position of radiator-valves..........................	58
" 17.—Remedying the air-binding of box-coils.......................	59
" 18.—How to use stove as hot-water heater.........................	60
" 19.—Relative value of pipe and steam coils........................	63
" 20, 21, 22, 23, AND 24.—Warming churches (plan of placing a coil in each pew)....................................65, 67, 68, 69	
" 25.—Steam-heating work, good and indifferent......................	70
" 26.—Piping adjacent buildings—pumps or steam traps..............	72
" 27 AND 28.—Advantages claimed for overhead piping...............	76
" 29.—Position of valves on steam-riser connections....................	78
" 30.—How to heat several buildings with a single apparatus...........	82
" 31 AND 32.—Air-binding in return steam-pipes......................	85
" 33.—Window-ventilators...	91
" 34.—Removing vapor from dye-house...............................	92

xiv   LIST OF ILLUSTRATIONS.

                                                                                                    PAGE

FIGURES 35, 36, 37, AND 38.—Ventilation of Cunard steamer "Umbria" (details).. 94, 96

" 39.—To prevent explosion of a steam-table....................... 104
" 40.—Cutting large nipples................................. 107
" 41.—Cutting crooked threads.................................... 109
" 42 AND 43.—Bending pipe.. 112
" 44.—Cutting various sizes of threads with a solid die................. 114
" 45, 46, 47, AND 48.—Contrivance for raising water in high buildings..116, 117, 119
" 49.—Heating water in large tanks............................... 124
" 50 AND 51.—Heating water for large institution.................125, 126
" 52.—Faulty elevator-pump connections.......................... 129
" 53.—Means of clarifying water for house-boilers.................... 131
" 54.—Differential ram for testing fittings.......................... 135
" 55.—Automatic pump-governor.................................. 137
" 56.—Cast-iron safe for steam-radiators........................... 138
" 57, 58, 59, 60, AND 61.—Methods of graduating radiator-surface according to the weather.................140, 141, 142, 143, 144
" 62 AND 63.—Methods of preventing the fall of spray from steam exhaust-pipes......................................145, 146
" 64.—Steam-heating apparatus and plenum system in the Kalamazoo Insane Asylum (a detail)............................... 147
" 65 AND 66.—Heating and ventilating a prison (details).............149, 150
" 67.—A form of expansion-joint................................. 153
" 68.—Low-pressure hot-water system for heating buildings in England.. 157
" 69, 70, 71, 72, 73, 74, AND 75.—Steam-heating apparatus in the Manhattan Company's and Merchants' Bank Building, New York (details)...............163, 165, 166, 168, 169, 170, 171
" 76, 77, AND 78.—The boilers in Manhattan Company's and Merchants' Bank Building, New York (details)..................173, 174, 175
" 79 AND 80.—Steam-heating apparatus in the Mutual Life Insurance Company's Building, New York........................179, 181

LIST OF ILLUSTRATIONS.   XV

                                                                 PAGE
FIGURES 81, 82, 83, AND 84.—The setting of boilers in Tribune Building, New
    York..183, 184, 185, 186
"   85, 86, 87, AND 88.—Warming and ventilation of the West Presbyterian
    Church, New York (details)...................188, 189, 190, 191
"   89 AND 90.—Heating-apparatus, Fine Art Exhibition Building,
    Copenhagen 193, 194
"   91, 92, 93, AND 94.—Warming and ventilation of the Opera House,
    Ogdensburg, N. Y..........................195, 196, 197, 198
"   95, 96, 97, AND 98.—Systems of heating houses in Germany and
    Austria......................................199, 200, 202
"   99.—Arrangement of system of steam-pipes in streets of New York
    City.. 204
"   100, 101, 102, 103, 104, 105, 106, 107, AND 108.—Some details of steam
    and ventilating apparatus as used on the Continent..206, 207,
                                                        208, 209, 210, 211
"   109.—Applying traps to gravity steam-apparatus..................   212

BOILERS.

BLOWING OFF AND FILLING BOILERS.

Q. WILL you please reply to the following inquiry? It is a question of steam-heating. The boiler is a tubular boiler. My man blew the boiler off with ten pounds of steam on. Would it be better to fill it with water again during the summer season, or not?

A. It is reasonable to presume that when your man blew the boiler off with a pressure of ten pounds of steam in it the walls and furnace were still quite hot. This is not considered good practice, as the heat of the fire-bricks, etc., is sufficient to heat and expand the under side of the shell of the boiler, when there is no water in it, to an extent that is considered detrimental to the boiler-plates. To avoid this possibility when there is more than one boiler, the steam-pressure from one of them may be let into any of the others that has been without fire for several hours, and the water and scum, etc., ejected in that way.

In the case of a single boiler, when it is going to be put out of commission, fill it to "*three cocks*" and blow down to "*one cock*" repeatedly, and until you think, from a previous knowledge of the condition of your boiler, that it is clean, or that you have removed all that can be removed in this manner. Then let the fires out and next day run off the water by gravity, when the boiler should be opened and the remaining matter removed as thoroughly as possible. A very good way, then, is to nearly fill the boiler with water, after which a few gallons of crude oil may be put in, and the boiler entirely filled with water. Then draw off all the water *slowly* from the boiler, and the oil will be brought in contact with every portion of the inside of the boiler.

Previous to starting up again in the fall of the year, fill up the boiler and wash it out, and should any scales be loosened in the mean time, remove them.

The above method is better than filling with water, but filling with water is better than taking out the handhole and manhole plates and allowing air to pass through a damp boiler.

Some open their boilers and dry them out thoroughly with heat, and try to preserve them dry for the summer, but the boiler is usually cold enough to condense moisture from the atmosphere, and rust. Again, if the boiler is kept perfectly dry and closed, the scale will not loosen from the tubes, if any is present.

WHERE A TEST-GAUGE SHOULD BE APPLIED TO A BOILER.

Q. PART of "Rule 40" of the General Rules and Regulations of the Supervising Inspectors for Steam-Vessels reads : "In applying the hydrostatic test to boilers with a steam-chimney, the test-gauge should be applied to the water-line of such boilers."

What we desire to know is, why the gauge would not be just as well screwed into the head of the steam-chimney or dome as at the water-line ? What real difference can it make ?

A. The object of the provision of the rule above stated is, to provide that the boiler will be subjected at all its parts to as near as possible the pressure required by law. When a boiler is just filled with water there is a pressure at its bottom of one pound for every twenty-seven inches of the height the boiler may be, and no pressure at its top. If, then, a boiler is twenty-two and one-half feet from the water bottom to the top of the steam-chimney—not an unusual thing in marine boilers—there will be a pressure of ten pounds per square inch on a gauge at the bottom of the boiler before the pump is at all applied. If thereafter the pressure is applied, and this gauge registers forty pounds, there is but a pressure of thirty pounds at the top of the dome. On the other hand, if the gauge were attached to the highest point, and fixed at that level, there would be actually fifty pounds per square inch at the bottom. For this reason the water-line is taken as the best position for the gauge.

It is not absolutely necessary that a hole should be made in the boiler at this position, but the gauge must occupy that level, and the gauge-pipe be full of water, that the head of water within it may act with or against the spring of the gauge, as the case may be.

DOMES ON BOILERS.

Q. I have noticed a recent specification for boilers, very minute in detail.

First—I cannot see why a single-sheet standard-steel boiler (such as is made by the Erie City Iron-Works and others) was not demanded in it, in place of one with so many useless seams.

Second—Why do they call for "domes" on the boilers? In the country we do not see the use of cutting a hole in a good strong boiler to rivet a piece on.

I wish you would provoke a general discussion of *domes*. I think a good deal of money is expended annually on these useless relics of a bygone theory.

A. The boilers in question were 66 inches in diameter by 17 feet 9 inches in length. To make one of these from one sheet of steel would require a sheet 18′9″x17′6″. The writer of the specification in question says he has yet to hear of the making of a sheet of steel of that size. He also informs us that he is of the opinion there is no machinery in the country fit to roll such a sheet, and that if there were, boiler-makers are not provided with the bending-rollers of sufficient length to make it into a cylinder. The largest boiler made by the works you mention, that has come to our knowledge, was 60 inches in diameter by 16 feet long, and was made of two sheets of steel, as we believe all the boilers are that have been made in this manner, and which, of course, necessitates *two* longitudinal seams. With these boilers the fibre of the steel—and we cannot entirely ignore the fact of there being a fibre in rolled steel—runs with the length of the boiler, and the bending is done in the opposite direction, both factors which militate against the strength of a cylinder in the direction where it requires the greatest strength.

The boilers you criticise were made, we are informed, of the best grade of American iron in the market. The sheets were long enough to go around in the direction of the length of the fibre, and there was but one longitudinal seam in each course, and therefore one less longitudinal seam than in the boilers you advocate, and the seam was alternately on opposite upper quarters of the boiler. Five sheets or courses were used in the length of the boiler, and these had to be joined by seams, but as the metal of a cylindrical boiler is loaded but *one-half* in the direction of its length that it is in the direction of its circumference, the factor for safety at circumferential seams (in courses and heads) is much greater than in the longitudinal seam.

The argument may be made for mild steel that it is homogeneous throughout, and that there are no hard spots in it; but steel from different makers presents such vast differences in quality and degrees of hardness or ductility that some engineers still prefer to use fine brands of iron to taking any chances with steels, when they cannot designate the makers.

With regard to the question of domes, there is some difference of opinion, with the preponderance in favor of having domes or drums. It is admitted a dome does not strengthen a boiler, but for heating apparatus, where large main pipes are used, they are almost indispensable. Who can expect to attach a 5-inch pipe directly to the shell of a boiler, and 15 inches or so above the water-line, and draw steam directly into the pipe without producing a *waterspout?* There are cases in the writer's mind where the water went from boilers so fast from this cause that he was in doubt whether it was best to run from the boiler-room or climb on top of the boiler and "throttle her down." *

Of course, "dry-pipes" and "deflecting sheets," and such contrivances, will obviate all this to some extent, but what is the use of producing a condition for the sake of applying a remedy, when with a dome properly put on the remaining strength of the boiler is as great as the strength of the longitudinal seam?

---

* "Throttling down" means partly closing the main steam-valve.

EXPANSION OF WATER IN BOILERS.

Q. I HAVE a number of horizontal boilers, 48 inches in diameter by 16 feet long, in which the increase in bulk of water is very apparent when the boilers are first warmed up. What I desire to know is, should I fill a boiler with water at a temperature of say between 40° and 50° Fah. to "two cocks," or, in other words, to the second gauge, how may I calculate how much higher the water will be, due to expansion alone, when steam is up to 69 pounds pressure?

A. You have not sent us sufficient data to give anything like an accurate reply. In fact, we do not see how we could give a fair approximation without a drawing of your boiler-heads, showing the size and number of tubes and the positions of the water-gauges. When the bulk of water at 40° Fah. is 1,000, the bulk at 307° (60 pounds of steam) will be 1,090. But from this must be taken the quantity of water which has been made into steam to fill the steam-space of the boiler; and the enlargement of the boiler itself, due to the increased heat, will enter into the problem, though not to a very great extent.

MUZZLES OF MAGAZINES OF HOUSE-HEATING BOILERS.

Q. WILL you inform me if there is anything better than cast-iron with which to form the muzzles of magazines for house-heating boilers? I have tried wrought-iron and cast-iron in a base-burning boiler of my own design, and neither last more than one winter.

The cast-iron muzzles spread at the lower end, and pieces fall out of them, and the wrought-iron spreads so that it almost turns up at the outer edge. Is there any remedy for this?

I use strong draught and a bright and small fire, as I imagine this is the proper way to burn the coal to obtain the best results in point of economy of fuel.

Any points on this subject that your experience can suggest will be thankfully received.

A. Cast-iron is presumably the best suited for magazine muzzles, but almost anything that projects downward into a hot fire, unless there is a water-circulation within it, will burn. The curling outward of a wrought-iron and the breaking off of parts of a cast-iron muzzle can be prevented if you arrange the edge of the muzzle so that it will be made up of a number of prongs set close together, or if slots somewhat like saw-cuts are made in its lower edge for a distance of two or three inches and one and a half or two inches asunder.

The reason the wrought-iron muzzle turns up is, that the lower edge of it is made very hot, while a few inches above it the body of the coal in the magazine and the conduction of heat into the body of the boiler leaves it comparatively cool. The hot lower edge is expanded and lengthened considerably in all directions, and must go outward on account of the cylindrical shape. This occurs with every considerable change in temperature, and gradually the hot lower end takes a permanent outward set, by being strained within itself, and presumably beyond its limit of elasticity, from which it cannot entirely recover when it cools. Each heating, therefore, is going to spread it larger and larger, until it is bell-mouthed. The same process goes on with cast-iron, except that, from its nature, it cracks and pieces fall from it before it turns far.

The slots will relieve this strain on the lower edge, and though it becomes equally as hot, the compensation of the prongs as they widen into the slots prevents spreading, as each prong is an independent piece. This will not prevent the burning of the prongs in a hot fire, but they will burn backward slowly.

PIPE-CONNECTIONS TO BOILERS.

At a meeting of the Executive Committee of the Manchester Steam Users' Association, held May 30, 1885, Mr. Leavington E. Fletcher, Chief Engineer, presented his report, from which it appears that an inspector of the association, in making an examination, found that the safety-valve would not blow when relieved of weight, although there were ten pounds of steam on the boiler. At a pressure of thirty-nine pounds it suddenly began to blow violently, and subsequent investigation developed the fact that three weeks previous a new rubber joint had been made under the safety-valve, but that the person in making the joint did not cut the centre out of the gasket, and that it did not burst until the pressure mentioned was attained. He cites another case where two men were killed in Liverpool by the bursting of a feed-water heater, in consequence of a rubber gasket not being cut out. The water was forced into the heater from the pump, and the pipe that was stopped was between the heater and the boiler. The pressure was sufficient to burst the heater-tank.

He also refers to a growing practice of using malleable-iron safety-valve levers. These levers are nothing but cast-iron made malleable, and may be defective through insufficient or improper manufacture, and it is certainly not a fit material for safety-valve levers. A case in point was the sudden breaking of one of these levers that looked like wrought-iron, deluging the boiler-room with steam and water.

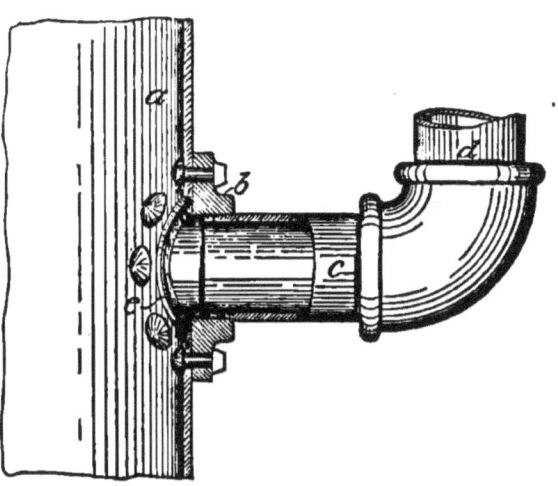

FIGURE 1.

[In New York and the big cities, at the present time, nearly all flanges are riveted to the domes and shells of the boilers, and the safety-valves attached by screwed nipples ; no soft joint being used inside the safety-valves or stop-valves. This prevents the necessity of a gasket between these valves and the boiler, and has been brought about through the annoyance of having to renew soft joints annually or oftener in hotels, apartments, and office buildings.

If, instead of using India rubber, with its web of canvas or fibrous materials, asbestos cardboard were largely used for packing flanges, the dangers cited from the neglect to cut out the rubber would be overcome in consequence of the cardboard being softened to a pulp by the moisture of the steam or water, and at best having little sustaining power, being about of the strength of blotting-paper of the same thickness.

The method shown in Figure 1 is that now almost universally used in boilers for our large cities and by boiler-makers accustomed to doing that class of work.—ED.]

PASSING BOILER-PIPES THROUGH WALLS.

Q. IN passing pipes through the brick walls of boilers I am forced to make holes much larger than the pipes, otherwise the pipes will rest on the bricks, should there be settling or heaving of the walls or boilers. Is there any practical method of accomplishing this without leaving large holes for the passage of cold air to the boiler or furnace?

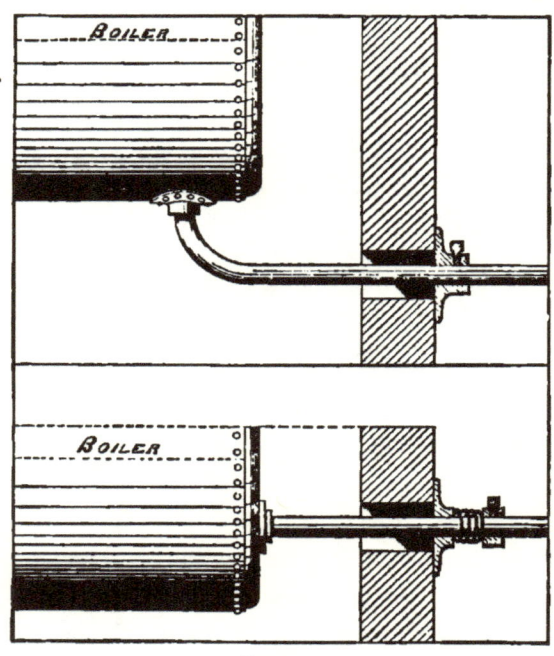

FIGURE 2.

A. Presumably the best method of accomplishing what you desire is to turn an arch over large pipes where they pass through the walls, leaving sufficient clearance — one or one and one-half inches; then fasten a flange to the pipe with a set-screw as shown, the flange being large enough to cover the hole. With small pipes an arch is not necessary. This provides for movements of the pipe in the direction of the plane of the wall.

If the pipes move in and out slightly, caused by expansion, and it is desirable to keep tight joints, use a collar on the pipe and a loose flange with a spiral spring between them, as shown in the lower part of Figure 2, and the difficulty will be obviated.

SUFFOCATION OF WORKMEN IN BOILERS.

A CORRESPONDENT writes:

Sir: I noticed the following in the *Locomotive* for January, 1885:

"Considerable comment has been made in some of the newspapers over the death of the engineer at the Laflin & Rand Powder-Works. It is asserted that he was overcome by carbonic-acid gas in the boiler; but we do not see how the boiler could become filled with this gas. We have entered boilers in about every imaginable condition, and we have never found it yet."

It is rare to find carbonic-acid gas in any considerable quantity in boilers; but should the manhole-plate of a boiler be removed overnight, and other boilers in the same boiler-room be fired for any considerable time, or "banked," with closed dampers, there is a possibility of the carbonic acid or carbonic oxide accumulating in the open boiler on the well-known principle of the former's precipitation into wells or holes in the ground. But an actual case of carbonic acid in a boiler, in which the writer was the principal actor, is as follows: A hole about thirteen inches in diameter had been made through the shell of the boiler into the dome. Through this the writer forced himself to the waist, with a candle and some tools to remove the "burs" from the edges of some small bolt-holes that had been made with a "cape" chisel and rounded with a drift-pin. After working a few moments the candle began to burn dim, and while looking for something to touch the wick with, it went out. This proved sufficient warning to one who had a smattering of physics, and he wiggled out of that hole in "less than no time."

The question may now be asked why the man did not "go out" as soon as the candle, but probably the reason lay in the fact that his nose was near the little holes, and the candle was at the lowest point in the dome. Both the man and the candle had vitiated the few cubic feet of air in the drum, and his body in the hole prevented the diffusion of the noxious gas into a greater body of air.

Very respectfully yours,   B.

FEEDING BOILERS.

Q. WILL you favor a novice at steam-fitting with an answer to the following problem? There is a manufacturing establishment in this town which uses six steam-boilers. These boilers are 27 feet long and 4 feet in diameter, and are in two gangs of three boilers each. The water, until this spring, was fed into the boilers by means of two mud-drums, H H, under the boilers, from a Blake double-acting pump, A, which has a water-cylinder of 6 inches diameter and suction-pipe of 5 inches. There were two branches from the pump, one for each gang of boilers. The heaters, B B, are close to the pump and about three feet from the boilers. The water, after leaving the heaters, went into the mud-drums through the pipes G′, under each gang. These mud-drums were 14 inches in diameter and 26 feet long, and were set at right angles to the boilers. They were connected to each boiler by pipes 10 inches long and 6 inches in diameter. This arrangement always gave good satisfaction, so far as the feed was concerned. The mud-drums showed signs of age this spring, and the company decided to do away with them. They did so. The holes in the bottoms of the boilers were closed up, and a hole cut in the back end of each boiler 3 inches in diameter. A 2-inch pipe, G, was led from each heater around to the back of the boilers, and was then connected with a 3-inch pipe, C, on each gang. These 3-inch pipes have tees in them opposite each boiler, and a 3-inch pipe, C′, from the tees to each boiler.

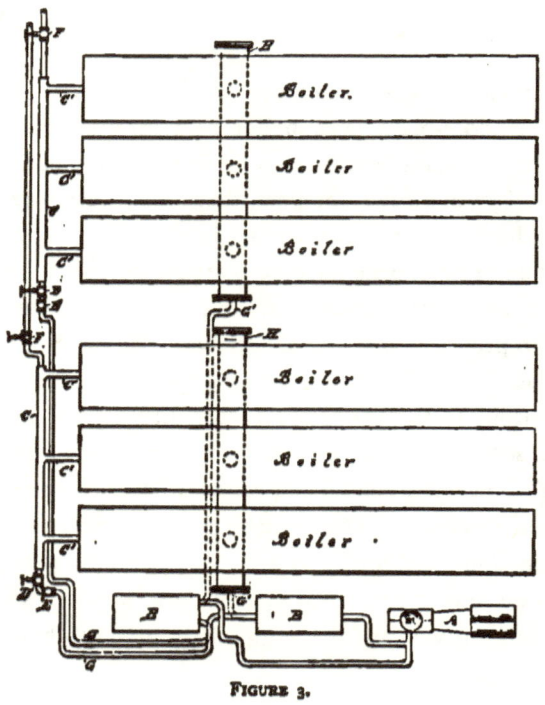

FIGURE 3.

This arrangement materially increased the length of pipe. The furthest gang has now 60 feet of pipe, without including the heater, which has seven 2-inch pipes 7½ feet long.

In pumping up the boilers after repairing, everything worked like a charm; but when steam was got up we could get no water through the pipes at all, except by running the pump at such a high rate of speed as to threaten the destruction of it in a very short time.

At first we thought it was the pump. It was taken apart and found to be in first-class condition. It had been to the shop this spring, bored out, and put in good shape.

How can you explain what is the matter? The company owns three establishments of this kind, situated in different parts of the country, and the engineer of one of them has a sort of supervision over the whole. It was by his instructions that the change was made, and now, when it doesn't work, he condemns the pump, but cannot find anything the matter with it. Now (1) what difference would it make to have the pipes from the tees in feed-pipe to the boiler reduced to two inches? This would give a 2-inch feed into a 3-inch and 2-inch to the boilers. (2) What would be the pressure on the pump if the feed was put into the mud-drums again? The boilers carry a working pressure of sixty pounds.

I hope I have made this plain for you to understand what I mean, and, in case I have not, I inclose a diagram showing the position of boilers, etc., and the change in the pipes.

A. 1. It would do no good to reduce the pipes C′. The size of the pipes have nothing to do with pressure, except in so far as the resistance to the flow is increased or decreased with the diameter of the pipe.

2. The pressure on the water-end of the pump is the same per square inch as the steam carried on the boiler, plus the weight of a column of water equal to the difference of level of the water-line of the boiler and the pump; added to which must be the resistance to the flow of the water through the feed-pipes and heaters when you are pumping. This resistance is generally an unknown quantity, and increases in a ratio about as the square of the velocity of the flow, and directly as the length of the pipes when they are straight or nicely curved, but in a much more rapid ratio in ordinary screwed pipes and fittings.

In your case, as the pipes are now fitted, the resistance to the flow is enormously increased over the old method, and presumably you have reached a point where the leakage around the piston and backward through the valves is such that only by quick running can sufficient water be forced through the pipes.

You give no data by which we can find the quantity of water used per hour; but assuming each boiler to be 60-horse-power, the quantity of water used per hour cannot be short of 14,400 pounds, and probably reaches 16,200 pounds per hour. As the pumps for any boilers should be capable of adding four times as much water as the boilers can evaporate in a given time, so as to be able to "catch up" should they be stopped for a time, and for other obvious reasons, it will in your case require a pump capable of adding water to the boilers at the rate of, say, 60,000 pounds per hour, or 1,000 pounds per minute, to do which the velocity through a 2-inch pipe at the pump will be 12 feet per second, and the pump will have to run at the rate of 120 strokes to the minute, if the stroke of the pump is 8 inches, making no allowance for leakage under valves, clearance, or loss in any way, which may bring it up to 150 strokes per minute.

The forcing-pipes, G, are too small in diameter, as the velocity per second through a single one of them will be 6 feet. *Two* feet per second is a fair velocity through a feed-pipe and its valves and bends.

BOILER-LUG.

THE illustration, Figure 4, represents a horizontal boiler-*lug* which is separable.

The shoe *a* is riveted to the boiler-shell before the tubes are put in, so as to admit of having the "point," or driven end, of the rivet on the inside of the shell. The bracket *d* slips into the shoe, forming the "*lug*."

We are informed that the driving of the rivets upon the inside insures tightness, as it admits of the wrought-iron or steel of the boiler being drawn and hammered tightly against the casting, whereas if the rivets were driven on the outside no *drawing* could be done with the hammers, the shaping of the rivet only being all that is possible, which will generally leave a leakage past the head of the rivet and out between the metals of the plates.

The object of the arrangement is to admit of passing a boiler

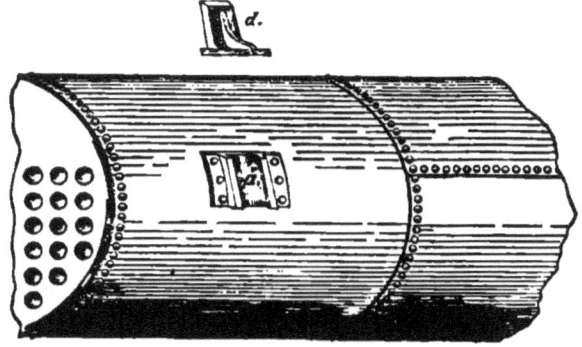

FIGURE 4.

through a doorway through which it would not pass if the lugs were made in the ordinary way.

W. J. Baldwin, in his work on steam-heating, page 67, recommends these boiler-lugs, and points out the objections to the method of riveting lugs to a boiler after the tubes are in, or to bolting them on.

ISOLATING-VALVES.

THE *Mechanical World and Steam-Users' Journal* of December 24, 1884, gives the illustrations Figures 5 and 6, which represent a valve to be used in the steam-supply pipes from boilers when two or more boilers are to be connected with one system of distribution. That journal refers to the pleasure it affords it to bring the important point of the isolation of steam-boilers before its readers, says the question to boiler-users is one of the first rank, and remarks in substance:

"When two or more boilers are worked together they are connected in the steam-outlets and become one machine. A branch-pipe containing the stop-valve from each boiler enters the main steam-pipe, and when the valves are all open there is an equilibrium of pressure throughout the whole range, however much or little either of the boilers may be fired. Now, conditions arise which can never exist in a single boiler. When only one boiler is used the various appliances upon it can be affected by the water or steam contained therein, and as these appliances are made of ample size for the single boiler no danger is incurred. But when there is a range of boilers connected in their steam-ways, then any one of these boilers may be placed so that its appliances may have to do duty for the whole range, and therefore prove inadequate. It is customary to apply a check-valve for the water-feed, to prevent such water, after it has entered the boiler, from being driven back again to other boilers, and what is wanted is a similar check-valve for the steam—a valve that will permit the exit of steam from the boiler, but not the inlet. If the pressure in the main steam-pipe rises an almost infinitesimal amount above that in one of the boilers, that boiler should be automatically and certainly shut off. A great number of attempts have, we believe, been

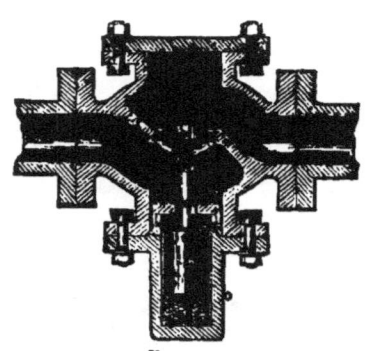

FIGURE 5.

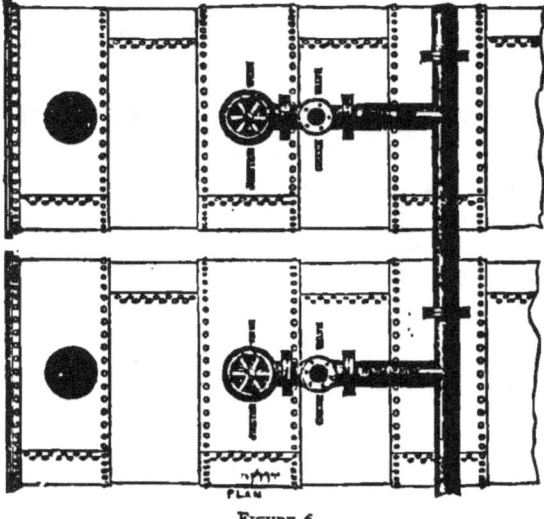

FIGURE 6.
THE ENGLISH VALVE.

made to achieve this result, but none successfully until the arrangement brought out some time ago by Messrs. J. Hopkinson & Co., of Huddersfield. This appliance and its application we illustrate herewith. Figure 5 is a section, and Figure 6 shows the mode of attachment to boilers. Referring to Figure 5, A is a brass valve closing upward; it is connected to an iron float or plunger submerged in mercury contained in the cylinder C. The inlet of steam is from the right hand. The iron float is so made that it has a determinate amount of buoyancy given to it; the amount usually adopted by the makers is one-quarter of a pound to the square inch. When, therefore, the pressure of the inlet is one-quarter of a pound per square inch above that in the main steam-pipe the valve is pressed downward, and the steam flows out; if, however, the pressure in the boiler and main pipe is equal, or that in the boiler is less than that in the pipes, the valve floats up

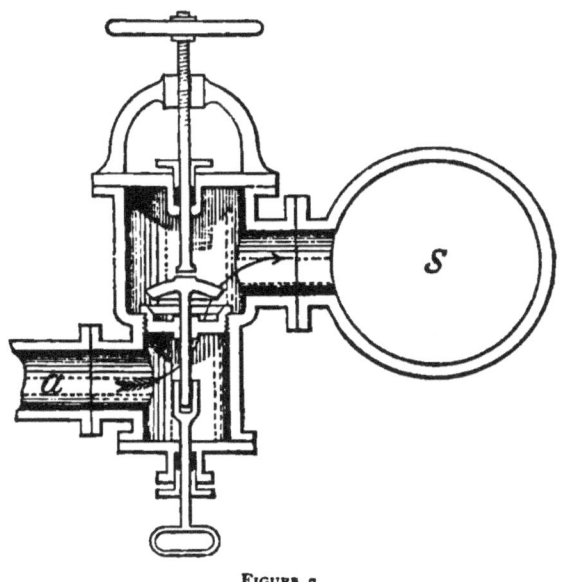

FIGURE 7.
THE AMERICAN VALVE.

and is closed. If the valve were loose in its spindle it would, in passing the steam, strike against its seat violently and continuously, but in this invention this action does not take place. It is designed to close gently and without the least impact, because while the float moves in mercury contained in the vessel C, practically without friction, yet it only moves at a comparatively slow rate, because the speed of transference of the mercury through the annular space between the float and cylinder is limited to a fixed amount. This valve is an admirable addition to a range of boilers that are fitted with low-water safety-valves. Should any of the boilers become low in water during the night, or from any cause, the check-valve will prevent steam from flowing in from the other boilers, and the low-water safety-valve will only have to discharge the steam from its own boiler, and not from the range."

The claims made for a reliable valve of this kind are :

"It will completely isolate any one boiler of a series, and, if such boiler be empty, will act, without requiring the vigilant watchfulness of any one, as a self-acting stop-valve. It will be a protection against a leaky stop-valve, or in the case of any one mischievously or unwittingly opening the stop-valve when the boiler is off, and when some person perhaps is inside cleaning the boiler or otherwise engaged. It will obviate the risk to which boiler inspectors have been subjected, and entirely relieve them from the constant dread which has haunted them in the prosecution of their duties. It will be a protection against loss of steam from all the boilers in the case of any one of a set becoming leaky or any of the pipes breaking (other than the main pipe), as it renders the boilers separate and distinct from the others under such conditions, while for all normal and required purposes it is no inconvenience whatever, as it permits the steam to flow in the ordinary and appropriate channels, as if no steam check-valve had been applied."

While not wishing to question the enterprise and energy of Messrs. J. Hopkinson & Co., we are forced to take exception to the statement that "a great number of attempts have been made * * * to achieve this result, but none successfully" until this arrangement, for on page 172 of the *Sanitary Engineer* of January 25, 1883, a valve of this description was illustrated and described as in successful operation in the plant of the New York Steam Company, at "Station B," on Greenwich Street, in this city, where we suppose the largest battery of boilers in the world is in operation day and night, supplying a large part of the lower business portion of New York with steam for power and warming purposes. The full extent of the plant is sixty-four boilers, each of 250-horse-power, of the water-tube pattern of Babcock & Wilcox, aggregating 16,000 horse-power, the greater part of which are in use. This valve we reproduce—Figure 7—and we have been informed by the engineer-in-chief (Mr. Charles E. Emery) that the valves have already proved their usefulness and silent working, as in the case of a split header in one of the boilers, when no one could approach it until it had blown off through the break, when the valve isolated the single boiler from the others, and prevented the escape of the immense accumulation of steam in miles of pipe.

Who the inventor is we do not know, but to Mr. Charles E. Emery, M. E. and C. E., Chief Engineer of the New York Steam Company, belongs the credit of its first application in the United States.

THE EFFECT OF OIL IN BOILERS.

The following article we take from the *Locomotive*, to show the effect of at least one quality of oil in boilers, which is generally understood to be a mineral oil. About the danger of animal oils in boilers there appears to be not the least question, but many do not hesitate to use large quantities of cheap mineral oils for boiler-purging, not knowing or not considering that many of these cheap oils are little better than a residuum of some other manufactured products, and containing many, if not all, of the heavy constituents of crude petroleum, and, for all the public may know, other substances which have been added in the manufacture for the purpose of giving "body" to the oils, or to increase their efficiency as lubricants. As so many of our readers are interested in the use, construction, and maintenance of boilers, we print nearly in full:

"The illustration gives a better idea of the effect produced than pages of verbal description possibly could. It is from a photograph, and is in no wise exaggerated.

"The boiler from which the plate shown in the cut was taken was a nearly new one. It was made of a well-known brand of mild steel, and that it was admirably adapted to the purposes for which it was used was proved by its stretching as it did without rupture. The dimensions of bulge shown are four feet lengthwise of the boiler, three feet girthwise, and nine inches deep. The metal, originally five-sixteenths of an inch thick, drew down to one-eighth of an inch in thickness at the lowest point of the 'bag' without the slightest indication of fracture.

"The circumstances under which the bulge occurred may best be described in the words of the inspector who examined the boiler, and are as follows:

"'Last Tuesday morning I was called in great haste to the ——— Works. Upon arrival I found one of the boilers badly bulged, and with twenty pounds of steam up. I could give no explanation until I had thoroughly examined the internal parts of the boiler. I gave directions for cooling the boiler, and ordered top manhole-plate to be loosened, but not to be taken out until my arrival in the afternoon, that I might see everything undisturbed. This was done. On my arrival I took out the manhole-plates in top of shell and front head, * * * and made an examination.

FIGURE 8.

"'I found that the boiler had been cleaned the preceding Sunday, and at that time a gallon or more of black oil had been thrown into it. Monday morning the boiler was fired up, and was running through the day at a pressure of 90 pounds per square inch. At six o'clock Monday night the engines were stopped, the draughts were closed, and no more firing was done until nine o'clock. Upon going to fire up at this time the bulge was observed. From six to nine o'clock a pressure of only 40 pounds was carried.

"'Upon examination I found the entire boiler saturated with this oil.'

"This is almost certain to be the result of putting grease into a steam-boiler. It settles down on the fire-sheets when the draught is closed, and the circulation of water nearly stops, and prevents contact between the plates and the water. As a consequence, the plates over the fire become overheated, and under such circumstances a very slight steam-pressure is sufficient to bag the sheets. Unless the boiler is made of very good material the plate is apt to be fractured, and explosion is likely to occur.

"When oil is used to remove scale from steam-boilers, too much care cannot be exercised to make sure that it is free from grease or animal oil. Nothing but pure mineral oil should be used. Crude petroleum is one thing; black oil, which may mean almost anything, is very likely to be something quite different.

"The action of grease in a boiler is peculiar, but not more so than we might expect. It does not dissolve in the water, nor does it decompose; neither does it remain on top of the water, but it seems to form itself into what may be described as 'slugs,' which at first seem to be slightly lighter than the water, of just such a gravity, in fact, that the circulation of the water carries them about at will. After a short season of boiling, these 'slugs' or suspended drops seem to acquire a certain degree of 'stickiness,' so that when they come in contact with shell and flues of the boiler they begin to adhere thereto. Then under the action of heat they begin the process of 'varnishing' the interior of the boiler. *The thinnest possible coating of this varnish is sufficient to bring about overheating of the plates*, as we have found repeatedly in our experience. We emphasize the point that it is *not* necessary to have a coating of grease of any appreciable thickness to cause overheating and bagging of plates and leakage at seams."

IRON RIVETS IN STEEL BOILER-PLATES.

IN a paper read before the Institution of Naval Architects by J. G. Wildish, M. I. N. A., he points out that iron rivets in steel plates shear at a less pressure than the same rivets in iron plates, and goes on to say:

"Some further experiments were made at Pembroke, in 1878, with iron rivets in steel plates; $\frac{9}{16}$-inch and $\frac{3}{4}$-inch plates were used, made by the Landore Steel Company, the rivets for connecting the test-pieces, which were jointed with a double-riveted strap to represent the butts of outside plating, being $\frac{3}{4}$-inch and $\frac{7}{8}$-inch respectively. In some of the tests the countersinking of the holes was carried right through the plates, and in others to within one-sixteenth of an inch of the full thickness. This variation, however, gave no appreciable advantage either way; but from these experiments it appeared that the average single shearing stress of the $\frac{3}{4}$-inch iron rivet in steel plates was only 8.1 tons as compared with the 10 tons for the same rivet in iron plates. The mean single shearing stress of the $\frac{7}{8}$-inch rivet was $11\frac{1}{2}$ tons, which, after allowing for the difference in size, is a somewhat better result than just given for the $\frac{3}{4}$-inch rivet, but is still 2.1 tons less than for the same rivet in iron plates. This comparative weakness of the iron rivets when used for connecting steel plates was met by making the rivets larger, as well as by placing them closer together, but the larger rivets involved broader laps for the plating, thus objectionably increasing its weight as a whole.

"Precisely similar experiments to those just alluded to were made in 1880, except that steel rivets were used instead of iron. The results of these experiments were exceedingly uniform and satisfactory. They showed that $11\frac{1}{2}$ to $11\frac{3}{4}$ tons might be allowed for the single shear of $\frac{3}{4}$-inch steel rivet in steel plates, and $14\frac{3}{4}$ tons for a $\frac{7}{8}$-inch rivet. Great care was taken in the manufacture of the steel rivets; and to insure their being of uniformly good quality, a code of tests was prepared for guidance in making them."

[This is a matter that should be thoroughly considered by makers of mild-steel boilers in this country. Where the boilers are built under the specification and direction of an *engineer* the danger is not so great, as he either provides for steel rivets or decreases the pitch of the holes, so that the remaining steel of the plates and the strength of the iron rivets nearly balance; but those who still adhere to the old empirical rule—that twice the thickness of the plate equals the diameter of rivet and *three* times diameter of rivet equals the pitch of holes—and who have a set of templates that were laid out when they were young, but who are now building mild-steel boilers, would do well to get information on this subject, and if they do not consider the importance of the matter, the architects and steam engineers who do business with them should, as the intelligent boiler-maker charges no more for good work than the "don't know" or "don't care" one does for indifferent work. —ED.]

PROPORTIONS FOR RIVETS.

In the *Locomotive* for June, 1885, is given, with illustrations, what the inspectors of the Hartford Boiler Insurance Company have found by experience to be good proportions for hand-driven rivets for boiler-shells. In the proportions for plates ¼-inch thick the size of the rivet should be ⅝ of an inch in diameter by 1⅜ inches long. The rivet-holes should be from $\frac{21}{32}$ to $\frac{11}{16}$ of an inch in diameter. The diameter of the base of the conical head where it comes in contact with the plate should be 1¼ inches in diameter. The height of head, from base to apex of cone, should be ½ of an inch. To form this head, as well also as to furnish metal enough to properly fill the rivet-hole, the rivet must project ⅞ of an inch, before driving, beyond the plate. This is secured by using rivets 1⅜ inches long, as before stated.

In the proportions for $\frac{5}{16}$-inch plates, the rivet used should be $\frac{11}{16}$ of an inch in diameter. The diameter of the head should be 1⅜ inches, its height $\frac{9}{16}$ of an inch. The rivet-hole will be from $\frac{1}{32}$ to $\frac{1}{16}$ larger than the original diameter of the rivet-shank, or from $\frac{23}{32}$ to ¾ of an inch. To form this head and furnish metal enough to properly fill the rivet-hole it will be necessary to use a rivet 1⅝ inches in length.

In the proportion of rivets in ⅜-inch plates, the holes as punched will vary from $\frac{25}{32}$ to $\frac{13}{16}$ of an inch in diameter. The diameter of the base of the hand-driven head should be not less than 1½ inches, and its height ⅝ of an inch. To fill this hole and properly form the head it will be necessary to use a ¾-inch rivet 1⅞ inches in length.

For $\frac{7}{16}$-inch plates a $\frac{13}{16}$-inch rivet should be used, the hole as punched being from $\frac{27}{32}$ to ⅞ of an inch in diameter. Diameter of head at base 1⅝ inches, height $\frac{11}{16}$ inch. The length of rivet required will be two inches.

For ½-inch plates, which is the thickest that should ever be used for tubular-boiler shells, a ⅞-inch rivet, 2¼ inches long, will be found necessary to properly fill the hole, which will be generally $\frac{15}{16}$ of an inch in diameter, and form the head, which should be 1¾ inches in diameter and ¾ of an inch high.

The foregoing are the minimum lengths of rivets and sizes of heads which should be used for hand-riveting, and Figure 9 shows the proportion one-third full size.

The other figures show two excellent specimens of machine-driven rivets. The thickness of plate used in each case is ⅜ of an inch. In each the spread of the rivet or diameter of the head is 1½ inches, and

FIGURE 9.   FIGURE 10.   FIGURE 11.

the height in Figure 10 is ⅝ inch and in Figure 11 $\tfrac{9}{16}$ inch. To make these heads as shown, each rivet would necessarily be two inches in length by ¾ inch original diameter.

The centre figure (10) was received from the Baldwin Locomotive-Works, of Philadelphia, and the one on the right (11) from the Cunningham Iron-Works, of Boston.

WATER IN BOILERS.

Q. WHAT would be the condition of the water in a boiler used continuously, over and over, without any new water being added, the boiler being part of a gravity return-heating apparatus? Would it be dangerous?

A. We do not understand what danger you particularly refer to. The only salt deposited to any extent in the boiler would be carbonate of lime, for there would always be enough water in the boiler to hold the soluble salts in solution. The water returning from the radiators is practically distilled water, and contains no salts.

ACCIDENT WITH CONNECTED BOILERS.*

THE feed-tank explosion reminds me of an accident, or rather disaster, that came under my observation some time ago, arising, like the one you described, from simple oversight, and unfortunately attended, in like manner, with fatal consequences. As the cause of the disaster was somewhat unique, it may perhaps be of interest to your readers, and help to point the moral you drew from the explosion described by you—viz., the importance of constant care and vigilance on the part of those who have to do with the fixing of steam apparatus. I therefore venture to send you a description, and inclose a sketch in case you should think it well to give it a place in your columns.

There were two boilers set side by side, as shown in Figure 12, No. 1 being at rest, while No. 2 was at work. Between the two boilers there was an expansion-joint, the end of the pipe D working quite freely in the stuffing-box and gland at E. In order to effect some repairs to the stop-valve A on No. 1 boiler, the valve B was shut down and a

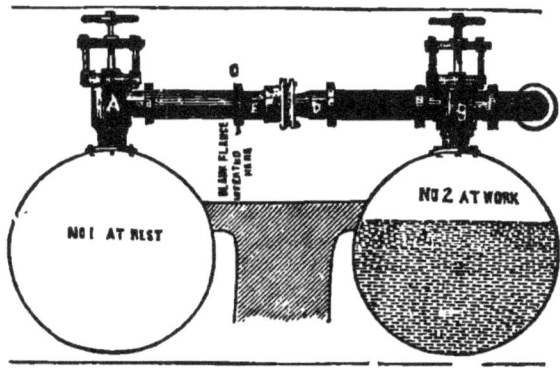

FIGURE 12.

blank flange inserted at the joint C. After this the valve B was opened again, and the mechanic commenced to take out the bolts securing the the valve A to No. 1 boiler, and had taken out two or three when the pressure of the steam acting on the blank flange at C shot the expansion-joint E along with the valve A right off the end of the pipe D, which was drawn out of the stuffing-box and gland at E, just like an arm is drawn out of a sleeve. In consequence of this the steam rushed out of No. 2 boiler through the open end of the pipe D, which was about six

---

* From an English exchange.

inches in diameter, and scalded the man to death, and also injured two others who happened to be in the firing place at the time. The man would appear to have been under the impression that the stuffing-box E was in some way connected to the pipe D, for it is almost incredible to think that a skilled mechanic would willfully commit such a blunder, and there can be little doubt that it was the result of a sort of easy-going careless indifference that sometimes prevails, and for which the poor fellow in this case paid a terrible price.

A SUPPOSED CASE OF CHARRING WOOD BY STEAM-PIPES.

Q. I HAVE a piece of wood about 6′ x 8′ x 3′ that was jammed in between the drum of a steam-boiler as a wedge to brace a partition, a sketch of which (Figure 13) I send.

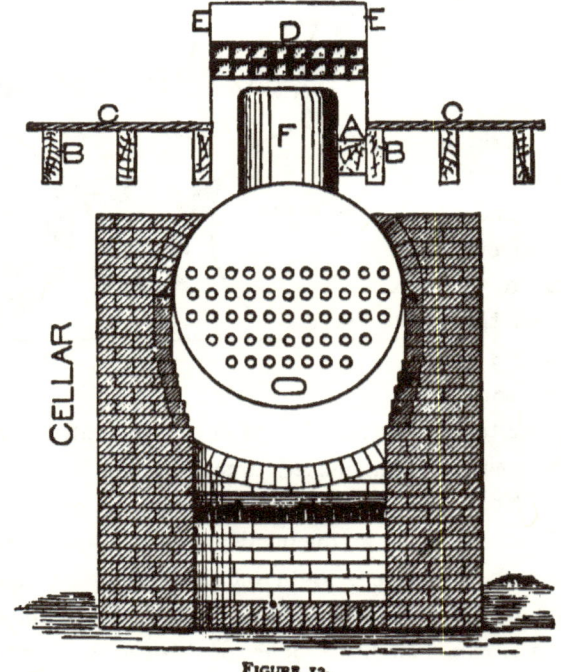

FIGURE 13.

A—Block of wood badly charred.
B—Beams carrying floor C, forming ceiling of boiler-room.
D—Windows in sides of caboose E, which is tightly built around drum of boiler F.

It is the worst case of charring by steam-heat I ever saw; in fact, I never saw anything like it before, and were I not quite sure of the facts, I would not believe it possible for the heat of steam to do the damage it shows. It had been in position between four and five months; steam carried between 50 and 60 lbs., for 12 hours per day. It was in such a position that there was no circulation of air around it. If you would like to see it, I will send it by express. I think it would do to write an article on it, and perhaps to illustrate. I may add that a beam with which it came in contact was charred a brown color and fell three inches away from the drum. The entire absence of circulation of air is my only explanation of the case. There was no oil nor anything of that kind near it; in fact, it was in a very awkward place to get at. I will try and sketch its position. Yours, J. W. H.

In a later letter Mr. H. writes:

"I send charred wood by express to-day. Have examined the place the wood came out of carefully again, with a trained practical scientist, and we found no cause to account for the charring other than the heat of the steam, unless the wood itself contained something."

A. We have had the wood carefully examined by one who has had much experience in this line, and he is of the opinion that although, to an ordinary observer, the wood has all the appearance of having been in a blaze at one time and quenched, it is a *bona-fide* case of charcoal forming, and that a spark has never formed on it.

Wood which takes fire and burns ordinarily never chars to any considerable depth ahead of the flame, and when the fire is extinguished the wood is found intact one-quarter of an inch from the surface of the blackened brand, but in this case the charring has penetrated two to three inches into the wedge, which is an indication of a steadily applied heat, but at a temperature too low to cause flame.

A temperature of 308° Fah. (the temperature of 60 lbs. of steam) is generally considered not sufficient to make charcoal from the most easily burned woods; but in the case of boilers and steam-pipes, too much dependence must not be placed on this supposition, as continued superheating of the steam by an improperly set boiler may be the means of giving much higher temperatures.

Assuming that a temperature due to 60 lbs. of steam at maximum density cannot char wood, steam can be superheated without materially increasing its pressure until it will make the pipe through which it passes red.

In the present instance the probabilities are that the boiler in question has been overheated—a thing not likely to be acknowledged by the engineer at this late date, even should the boiler show slight signs of it by leaking or by wrinkles in any of its sheets.

Our own impression is, that the temperatures at which woody carbon and its gases spring into flame are as fixed as the temperature at which water boils; but that condition which will produce this result may be brought about in ways not readily accounted for without knowing the minutest details of the existing conditions.

If cotton were a good conductor of heat, it would not ignite by spontaneous combustion, as a greasy cloth will prove if you lay it flat for an indefinite time; but make a ball of it, so the heat produced by the chemical reaction cannot pass off rapidly enough, and the result is heat and fire.

DOMESTIC BOILERS.

ABERDEEN, SCOTLAND.

Q. 1. How LONG will a properly galvanized-iron (hot water) cistern last? Why does this kind of cistern rust inside and the rust appear outside?

2. Is there any objection to a copper cistern tinned inside (for hot water)?

3. Can you recommend any other kind of cistern for a water-heating system?

A. 1. Our galvanized-iron domestic, or range boilers, as we call them in this country, have been in use about fifteen years. They do not rust on the inside sufficient to affect the color of the water, if in constant use, because they are closed from the atmosphere, and their outside appearance is always the same. They have come into general use because they are cheaper than copper tinned inside of equal strength.

2. If the price was not a consideration, we should prefer copper tinned inside. As we understand the practice in Great Britain, the hot-water tank, or boiler, as we call it, is placed alongside or near the cold

tank, and is not closed to the air. In such a case we should say that the galvanized-iron would soon rust and discolor the water.

3. In the large apartment-houses of this city tanks of ¼-inch or $\frac{5}{16}$-inch boiler-iron, from three to four feet in diameter by from six to ten feet in length, cylindrical in shape, with a manhole in one head, are used for warming water, the warming agent being generally the exhaust steam from the elevator engines, or pump for the hydraulic elevators, the thermal value of which is used this way. The sketch (Figure 14) shows how these tanks appear. Should the exhaust-pipe be three inches in diameter it is taken to one end of the tank and carried back and forth within it for a number of times, generally four, forming a coil; thence passes to the roof of the building, the condensed water being separated at the lowest corner before it ascends. These tanks are under pressure at all times. Water from a house-supply tank at the

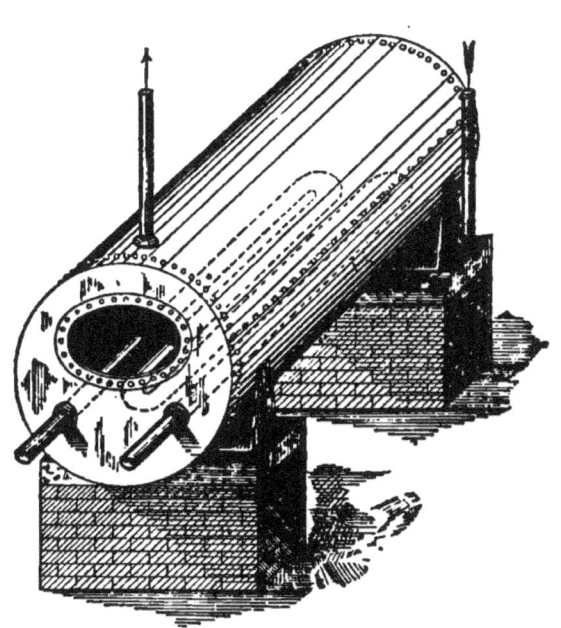

FIGURE 14.

roof is brought down into them, entering at the bottom and leaving again at the top, warmed, and of course rising within the building to an equal height with the house-tank. The warm supply from these tanks to the fixtures is sometimes fitted with a circulation-pipe for the purpose of keeping warm water constantly near the fixture. This pipe returns parallel to the rising pipe and enters the tank at the bottom. A good way is to carry a small pipe (three-eighths of an inch) from the head of these circulating lines to a height three or four feet above the house-tank, to act as an air-escape, and again to facilitate drawing the water from the line, should it be shut off for repairs, by allowing the air to draw in at the top. On pages 145 and 193 of the tenth volume of the *Sanitary Engineer* may be seen representations of domestic boilers and their connections.

VALUE OF HEATING SURFACES.

COMPUTING THE AMOUNT OF FUEL FOR WARMING BUILDINGS.

Q. If it is within the province of your esteemed journal to answer the following questions in your columns at an early date you will confer a favor on myself, and, I have no doubt, on other young men in the profession :

First—How am I to proceed to find approximately the amount of fuel necessary for a given duty in the warming of air of buildings when direct radiation alone is used?

Second—How am I to proceed when indirect radiation is used? Any data on the above questions will be thankfully received.

A. Accurate figures on these subjects are difficult to obtain, as the results must depend on the conditions found to exist in any particular building. When a building has large glass surfaces, or is composed of iron, or is unfurred—plastered on the outer hard walls—the air within it is cooled more rapidly than in ordinary buildings with average windows and other average conditions. This air is cooled by contact with the surfaces of the windows and walls, and these may be termed the "cooling surfaces" of the building. To counteract this cooling of the air, "heating-surfaces" must be added in the form of pipes or radiators.

The shape, position, and nature of the heating-surfaces are now factors, as well as the pressure of steam to be carried, in determining the proportion of surface for a given condition. It is here that the greatest difficulty presents itself, and the steam-engineer has nothing more than averages to depend on for his guidance. In the case of small corner rooms, with low-pressure steam, *one square foot* of average radiator-surface to each 50 cubic feet of air-space is generally considered ample, but this in itself is only a basis for an average.

With large rooms 1 to 60 is often enough, and in halls and stores 1 to 100 is usually considered sufficient, while many auditoriums and churches are found to contain a lower ratio still, 1 to 150 proving enough. After the heating-surface is determined for direct radiation the fuel used for a given time may then be approximately determined as follows: With low-pressure steam the condensation per square foot of radiator of usual shape, to maintain a temperature of 70° Fah., is about .25 of a pound of water per hour, and after the building is warmed with high-pressure steam it is likely to reach .33 of a pound of water per hour.

If, now, the rooms of your building to be warmed contain 1,000,000 cubic feet of air, and you have, say, 10,000 square feet of radiation surface in it, at a pressure not exceeding 5 pounds, you will have 10,000 × .25 = 2,500 as the number of pounds of water to be evaporated in the boilers or condensed in the coils and pipes in an hour. With good boiler practice it is then safe to assume that you can re-evaporate every ten pounds of this condensed water from the temperature at which it is likely to return to the boiler (about 200° Fah.) to steam again with one pound of coal. With this, then, you will have 2,500 pounds water ÷ 10 = 250 pounds of coal per hour as the least amount of fuel reasonably necessary for your 1,000,000 cubic feet of air when all the radiators are in use. Good practice may reduce this slightly, but it is just as likely to be exceeded.

In the case of indirect radiation the method of proceeding is somewhat different. There the condensation is more of an unknown quantity per square foot of radiator, and it will vary with the outside changes; but it admits of closer calculation when the amount of air to be passed into the building in a given time is known, for the reason that there can be no question then of the units of heat required, as all the air that enters must be warmed with an efficient apparatus, whereas with direct radiation no close estimate can be made of the units of heat applied to the air, other than by finding the amount of water condensed, or by assuming it from practical averages, though one practical authority assumes it to be one cubic foot of air cooled from the temperature of the room to that of the outside air in a minute for each square foot of glass, but he does not take into consideration the values of other cooling surfaces.

Again, take the case of a building with 1,000,000 cubic feet of air in it, and in which the air is to be changed *four times* in an hour. This gives you 4,000,000 cubic feet of air which is to be warmed from the temperature of the outside air to, say, 75° Fah. If, now, you assume the average mean temperature outside to be 35° Fah. for a given time, you will have to warm 4,000,000 cubic feet of air 40 degrees in each hour, which is equal to warming 160,000,000 cubic feet of air *one degree*. The heat given off by the cooling of one pound of water one degree may now be approximately taken as the warming of 50 cubic feet of air *one degree*, which will give 160,000,000 ÷ 50 = 3,200,000 the equivalent of 3,200,000 *heat units*. You may then assume the conversion of one pound of water at a temperature of the average return-water to steam at low pressures to be equal to 1,000 *heat units*. This, then, will give you 3,200,000 ÷ 1,000 = 3,200 pounds of water to be converted to steam, and, as before mentioned, if the boilers are capable of evaporating this hot water with one-tenth of its weight of coal, you will require just 320 pounds of coal per hour.

This gives approximately the fuel required, without taking into consideration the loss of heat in the main pipes, which should be fully covered by an addition of ten per cent.

COMPUTING AMOUNT OF RADIATOR-SURFACE FOR WARMING BUILDINGS BY HOT WATER.

CHATHAM, ONT.

Q. HAVING read your very instructive discussion on "Computing the Amount of Fuel for Warming Buildings," I have taken the liberty of troubling you for some information on heating by hot water. Our Government has just completed its new building here, and it is heated by hot water, and so successfully that a number of our citizens are thinking of warming their residences in the same manner. Direct radiation is the system adopted. As there is some diversity of opinion relative to the proportion of radiator-surface to cubic feet of air-space, I ask you for information, and would like to know what is the rule or average generally adopted, making allowance, of course, for exposed rooms. Taking steam as a standard at 1 to 50, and being the hotter of the two, would 1 to 35 be the proportion for hot water? Kindly give me all the particulars you can in reference to the system.

A. The discussion referred to is that immediately preceding. Hood says that the burning of one pound of coal may be safely estimated to add 7,000 heat units to the water in hot-water boilers. It is reasonable to assume that this will be so, as it is but one-half of the theoretical value of the average coal. He states, also, that when pipes four inches in diameter are 146.8° Fah. hotter than the air of the room the water contained in them cools exactly the equivalent of 1° Fah. per minute, if the temperature of the water is maintained constant. This, presumably, is the actual result for some particular condition or position of the pipes for direct radiation, and may be taken as an average.

From this he calculates the following table for 100 feet in length of the different sizes of pipes in general use, the quantities given in the table being pounds and tenths of a pound:

Table of the Quantity of Coal used per hour to heat 100 feet in length of pipe of different sizes.

Diameter of Pipe in Inches.	Difference between the Temperature of the Pipe and the Room in Degrees of Fahrenheit.														
	150	145	140	135	130	125	120	115	110	105	100	95	90	85	80
4	4.7	4.5	4.4	4.2	4.1	3.9	3.7	3.6	3.4	3.2	3.1	2.9	2.8	2.6	2.5
3	3.5	3.4	3.3	3.1	3.0	2.9	2.8	2.7	2.5	2.4	2.3	2.2	2.1	2.0	1.8
2	2.3	2.2	2.2	2.1	2.0	1.9	1.8	1.8	1.7	1.6	1.5	1.4	1.4	1.3	1.2
1	1.1	1.1	1.1	1.0	1.0	.9	.9	.9	.8	.8	.7	.7	.7	.6	.6

To find the quantity of hot-water pipe necessary for direct radiation proceed as follows: Take the difference in degrees Fahrenheit between the temperature at which you will be able to maintain your hot-water pipes and the temperature you wish to maintain the room at for a divisor, and the difference between the coldest outside weather and the temperature you wish to maintain your room at for a dividend, when the quotient will be the number of square feet, or fraction thereof, of surface of hot-water pipe to each square foot of glass-surface in the building. Thus: mean temperature of pipes 150° Fah. — temperature of room (70°) = 80°. Again, temperature of room, 70° — temperature outside (0°) = 70 ÷ 80 = .875. To this should be added from 30 per cent. to 60 per cent. for cooling of walls and cold air entering at doors and windows.

CALCULATING THE RADIATING-SURFACE FOR HEATING BUILDINGS—THE SAVING OF DOUBLE-GLAZING WINDOWS.

Q. PLEASE be kind enough to publish in an early issue of your valuable paper some simple rules for calculating the necessary amount of radiating-surface for heating buildings with steam or hot water. Please give rules according to the amount of glass, amount and quality of walls; also the difference between single and double glass.

A. W. J. Baldwin, in his "Hints to Steam-Fitters," says: "Divide the difference in temperature between that at which the room is to be kept and the coldest outside atmosphere by the difference between the temperature of the steam-pipes and that at which the room is to be kept, and the product will be the surface in square feet of plate or pipe surface for each square foot of glass, or its equivalent in wall-surface." This gives about *one* square foot of radiating-surface to each *two* square feet of glass for low-pressure steam. He also considers that from 7.5 to 10 square feet of ordinary outside wall cools as much air as a square foot of glass, or, say, we require one square foot of radiator to 15 square feet of outside wall. This does not provide for warming any outside air that may enter, and is seldom sufficient for ordinary practice. At least one-half more, or .75 of a square foot, is generally required.

With regard to the saving of heat by double glazing, General Meigs has pointed out that about one-third less heat is lost through two glasses placed with about one-fourth of an inch between them than through a single glass, but from this we must not assume that one-third less radiating-surface will do in such a room, as we must bear in mind that the radiating-surface is proportioned according to all the circumstances, walls, ventilation, etc., and that the heat saved is proportionate only to the number of square feet of radiating-surface necessary to counteract a given window area. For instance, if a room required seventy-five square feet of radiating-surface, although the windows had but sixty feet of glass-surface, ten square feet of radiating-surface would be the reduction, according to Baldwin's value for single glass.

In the neighborhood of New York, deductions based on the direct radiating-surface, compared to the cubic space, give averages with steam-pipes about as follows: Office-rooms, one square foot of radiating-surface to each 75 cubic feet of air-space; stores, one square foot to 100 cubic feet; lofts and upper stories, one square foot to 125 or 150; churches and large auditoriums, one square foot to 150 or 200.

The smaller a room is the greater the percentage of outside wall and window to the cubic contents. A room $10' \times 10' \times 10' = 1,000$ cubic feet may be a corner room and have two windows and two cold sides, and require about 25 square feet of surface. The next room to it may be $14'+2'' \times 14'+2'' \times 10'$ high $= 2,000$ square feet, very nearly, with one cold wall and two windows, and though it has double the cubic contents, it will require no larger radiator than the corner room.

Hood says that experiment has proven that each square foot of glass cools 1.28 cubic feet of air from the temperature of the room to the outside temperature in one minute. According to this, if we have 25 square feet in a window, with 70 degrees in the room and zero outside, we cool 1,920 cubic feet of air 70° in an hour, to maintain which we must condense very nearly three pounds of steam.

Experiments on radiators, such as are made in this country, give an average of three-tenths of a pound of steam condensed per hour to each square foot of surface, which would call for 10 square feet of radiator to the 25 square feet of window, making .40 of a square foot of radiator to each square foot of window-glass. This last rule gives a radiator-surface, as figured against glass, of twenty per cent. less than by Baldwin's method, and is probably too low for the ranges of temperature in this country.

AMOUNT OF HEATING-SURFACE REQUIRED IN HOT-WATER APPARATUS BOILERS AND IN STEAM-APPARATUS BOILERS.

Q. PLEASE state what amount of *heating* or *fire surface* is required in a hot-water boiler, as compared with the *heating* or *fire surface* in a low-pressure steam-boiler—the conditions of fire being the same.

That is, if in a low-pressure steam-boiler having a 30-inch fire-pot, experiment shows that 225 feet of heating-surface is ample, what should be the heating-surface of a hot-water boiler—the fire-pot being the same?

A. It depends entirely on the rapidity of the circulation in the hot-water apparatus.

In steam-boilers of good construction, a square foot of surface in one boiler has very nearly the same value as an equal amount of surface in any other steam-boiler, for the reason that the water is at liberty to circulate as rapidly as the force of ebullition can make it; but with a hot-water apparatus a square foot of boiler-surface has a value of depending entirely on the rapidity with which the water circulates in the pipes.

It is the water, as it passes over the surface of the boiler, which takes away the heat. When it passes rapidly it takes more, and less goes up the chimney. When the circulation is sluggish the heat is wasted up the chimney, no matter what the surface is.

CALCULATING THE AMOUNT OF RADIATING-SURFACE FOR A GIVEN ROOM.

MONTREAL.

Q. How many feet of 1-inch steam-pipe radiating-surface of the Nason form of radiator will it take to give a temperature of 140° Fah.; in a room which is 26 feet 6 inches long, 15 feet 4 inches wide, and 14 feet high, one side and one end exposed to the outer air, the other side and end abutting on well-warmed rooms, walls of 8-inch brick, lathed and plastered on the inside, three windows, each 7 feet 4 inches wide, fitted with double sashes? In the room is a plunge-bath 16′x8′x5′ sunk below the floor. This will usually be full of cold water; sides of tank and floor are tiled. Steam pressure on boiler 10 pounds, fitted to work as a gravity apparatus.

A. If you use long, flat radiators (single or double rows), and place them on the outside walls, 200 Nason radiator-tubes are about what will be required. There might be conditions which would raise the amount of surface to 250 square feet, but, again, with great precaution, such as drawing down window-shades and closing of the inside blinds, a test might be conducted which would give the required temperature with only 170 square feet. This does not take into consideration the effect caused by the plunge-bath, and we think nothing but experiment will determine the extra surface required. It may be that the evaporation from the surface of the bath will require much extra heat, or, if there is no ventilation, the air may become heavily charged with moisture, when a very small addition of surface will prove sufficient.

HOW MUCH HEATING-SURFACE WILL A STEAM-PIPE OF GIVEN SIZE SUPPLY?

Q. WILL you inform us as to the amount of heating-surface which a ¾-inch pipe will heat, with high-pressure steam? The pipe is to be carried about 500 feet in the ground, and is to be run through a wooden log or pipe. It drips from mill where we are to get our steam to the house. I suppose it will require a trap inside of the house where steam enters. What is the cheapest and best size to use? We intend to exhaust steam out through the house, and there are in the house 385 feet of heating-surface. What size and kind of trap is it best to use on the exhaust-pipe? Can we make a trap in the pipe itself and not use a steam-trap?

A. The loss of steam in small long pipes by friction and condensation is so uncertain, depending on local circumstances and conditions, that it would be hazardous to say for just how much surface a certain size and length of pipe will supply steam.

A ¾-inch pipe of the length you mention at 40 or 50 pounds pressure will supply steam for 385 square feet of surface with a very much reduced pressure at the heaters.

To insure success we should use 1¼-inch pipe for the work you mention. Use a steam-trap of intermittent action inside the walls of the house, to receive the condensation from the pipe before going to the heaters. If large piping is used within the house, so as to get a uniform pressure in the heaters, this same trap may receive the remainder of the water of condensation, but if very small piping is used another trap will be necessary.

COILS OR RADIATORS? ESTIMATING SIZE OF BOILER.

Q. I have a few questions to ask you. I want to heat a building, 50x100 feet. Would it heat as well by running the pipe in coils on the wall as by radiators in the rooms? Also, what size of boiler would be required and the best for that purpose? What size pipe should I start with and what size of return-pipe? Please tell me where I can get a good book on steam fitting and heating.

A. The same amount of pipe in long coils around the outside walls and under the windows will give a better result than radiators will. Use about one-fifth the surface in the boiler that you have in the coils, assuming that your coils are sufficient. Horizontal tubular boilers give a high efficiency when well set and will last a long time. There are also many fine water-tube and sectional boilers now before the public.

Baldwin's work on steam-heating is the best we know of.

CALCULATING AMOUNT OF HEATING-SURFACE.

Q. I have "Baldwin's Steam-Heating for Buildings," published by J. Wiley & Son, but the rule on pages 23 and 24 I do not quite understand. If not too much trouble, will you please make it a little more explicit? For example, publish the amount of heating surface and method of calculating on the following room, $19' \times 13\frac{1}{2}' \times 12'$ ceiling; lath and plastered on brick wall, with $1\frac{1}{2}$ furring between wall and laths. Three south windows, $3\frac{1}{2} \times 8$ feet, or 28 square feet to each window. Same sized room with plaster on brick wall.

A. You do not say how many outside walls there are to the room you mention; but as there are three windows, we assume the longest side (19 feet) to be the outside wall. The inner walls, ceilings, and floors you may omit on the assumption the other rooms of the building are warmed. Thus we have $19 \times 12 = 228$, less the window's area (84 square feet) $= 144$ square feet of outside wall, which, multiplied by 75 and divided by $1,000 = 11\frac{8}{10}$, or the equivalent of the wall-surface in square feet of glass for the lathed and plastered room, and in the same manner $144 \times 125 \div 1,000 = 18$ for the unlathed.

Thus we have in the first case $84 + 11\frac{8}{10} = 95\frac{8}{10} \times .5 = 47.9$ square feet of radiating-surface for the first room, and $84 + 18 = 102 \div \frac{1}{2} = 51$ square feet for the second room; to both of which should be added a generous allowance of surface to warm the air admitted for ventilation or leakage, if any.

COMPUTING COST OF STEAM FOR WARMING.

Q. CAN you give me the necessary information to calculate the amount of coal consumed in heating a room any given size by low-pressure steam? We heat our building on the low-pressure principle, and desire to rent heat to several of our tenants. What I desire to know is, how to calculate the quantity of coal it will require to heat, say, one square foot of radiator-surface every twenty-four hours, under ordinary conditions. By this rule I can decide how to charge for the different sized rooms, according to the amount of radiator-surface exposed.

A. A steam-radiator will condense from one-fifth to two-fifths of a pound of steam to water per square foot of surface per hour, varying with the nature of the surface, the kind and position of the pipes, and the temperature of the steam, in buildings where direct radiation only is used, and with the temperature of the air maintained at about 70° Fah. All other things being the same, a radiator will condense less steam at low pressures than it will at higher pressures, the increase of water of condensation being (about) directly as the temperature of the steam is in excess of the temperature of the air of the room. In other words, when the condensation with one pound pressure of steam at a temperature of 216° Fah. is represented by 216° — 70° = 146, the condensation at 50 pounds pressure will be represented by 298° — 70° = 228.

In the case of low-pressure steam, one-third of a pound of water per square foot of radiator is not too great an estimate for the man who has to sell the steam, especially when we consider that the condensation which takes place within the mains must be charged *pro rata* to each square foot of radiator-surface within the rooms.

According to this estimate, a radiator of 100 square feet of surface will condense 33 pounds of water, or about one-half a cubic foot, in an hour.

It is now necessary to consider at what cost you can evaporate water. A pound of coal will evaporate from six to twelve pounds of water, according to the way it is burned. Six pounds is poor practice; *twelve* presumably the very best possible. With these data before you, find the efficiency of your boiler, then take into consideration the cost of plant and interest on it, wear and tear, labor, etc., and you will be able to estimate within reasonable limits the cost of warming. To this, of course, add profit enough to cover contingencies.

The New York Steam Company charges sixty cents per 1,000 kals for the steam it sells, the kal being the equivalent of 1,000 heat units. This, in approximate numbers, may be taken as equal to the making of 1,000 pounds weight of steam from the temperature of water as it returns to the boiler from the heating apparatus, and on the assumption that one-third of a pound of water is formed per hour for each square foot of surface, the cost for a "100-pipe" radiator will be two cents per hour.

RADIATORS AND HEATERS.

A WOMAN'S METHOD OF REGULATING A RADIATOR.

INCLOSED I send you a sketch of a method invented or accidentally discovered by a woman for regulating and controlling the heat given off by a steam-radiator, which I think will be of interest to the readers of the *Sanitary Engineer*.

FIGURE 15.

No doubt you are aware that with a gravity apparatus, and, in fact, with almost all steam-heating apparatus, you must have all the steam turned on the radiator or it will fill with water and "pound"; that there can be no half-way about it, and that, consequently, in moderate winter weather a radiator-surface that is proportioned for very cold weather will be more than sufficient, and make the apartment too warm.

When the lady in question was told there was no way of modifying the heat but by shutting off all the steam, she suggested that the heater could "be covered with something to keep the heat in," and forthwith made what Figure 15 shows. She now draws the cover up and down to suit herself, making the radiator more or less effective as she uncovers or covers the pipe.

In summer-time she covers the radiator all up with her contrivance, which in color corresponds with the hangings of the room.

[We can see how by covering the pipes in this manner the air cannot come in contact with them as freely as when they are exposed, and that consequently as much water cannot be condensed or heat given off in a given time; but we would suggest to any one who would like to repeat the experiment that it would be well to use none but woolen goods—not that we are sure that cotton goods will take fire under such conditions, but to be on the safest side.—ED.]

IMPROPER POSITION OF RADIATOR-VALVES.

Q. INCLOSED I send you a sketch of a steam-radiator. At a glance a steam-fitter who knows his business will see that it is not properly connected, or at least that the valves are improperly attached to it.

I wish to call attention to an experience I had lately with a radiator similarly connected, evidently by a mistake or a blunder, as all the other radiators in the house (over fifty) were differently connected. In this building, which was an old one which was being fitted with steam-pipes, all the risers were exposed with the *tees* above the floors, so as to make direct and short connections. Globe-valves were used by the fitter in many places, though it was intended that all or nearly all should be angle radiator-valves. With this particular radiator it was found, after steam was up, that it remained full of hot water, or, if not full, with a great quantity of water in it, so much so that it always ran water from the air-valve when it was opened. Why this radiator should act so and none of the others give any trouble was not apparent, and the pipes were opened to look for a stoppage, when it was discovered that the valves were "wrong side to," as the steam-fitter calls it, and there was no stoppage. The valves were reversed, and afterward, though the stems of the valves were left in a vertical position, the radiator worked without rumbling of water.

The above are the facts. Perhaps you or some of your readers will explain why the changing of the valves should make such a difference, for, if I am informed rightly, steam-fitters turn the valves the other way—not to make the radiator work properly, but to make it convenient to pack both valves without shutting steam off from the whole house, by simply closing both and waiting for the steam within the radiator to condense.

A. We represent in our drawing (Figure 16) water in the base of the radiator. As there is practically as much pressure in the return end of a radiator, in an ordinary system, as there is in the steam end, the two globe-valves will "trap" the base of the heater full of water, as represented. When condensation takes place, as it must within the pipes of the radiator, the greater pressures in the pipes will force inward at the two ends and nearly alike. This will prevent the water which is accumulated in the base above the valve line from flowing out easily, as it should, and will make it assume a level still higher than we show. This will make the water rise against the bottom of the tubes and allow it to pass up in the tube on which is the air-valve, when the latter is opened, relieving the pressure in the heater still more. The natural result is the partial filling of the radiator with water, the accumulating column of which preponderates against the entering steam at the lowest or return end intermittently, thus finding its way out of the heater sufficiently not to let the heater become cold, but accompanied by the noise mentioned.

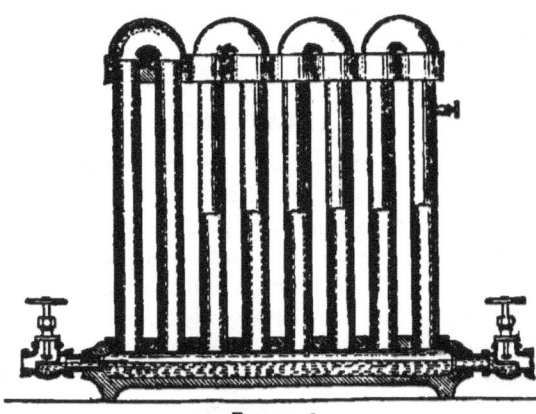

FIGURE 16.

If the valves are turned the conditions are somewhat better, as then the steam has a clear passage, though if the globe-valves must be used it is better to place them with the stems sidewise, but not quite down to the level. This gives a clear and level waterway on a vertical section.

HOT-WATER RADIATOR.

ST. JOHNS, N. B.

Q. Do you know of a hot-water radiator for use in private dwellings, in place of box-coils for house-heating by hot water? If so, kindly inform us.

Hot-water heating is largely used in our city, and has proved highly satisfactory and very economical of fuel. We construct a low-pressure hot-water heating apparatus, open to the atmosphere, for private dwellings, and must say we prefer it to steam for that purpose, although we put in low-pressure steam gravity apparatus, working at one pound pressure, also. But the universal opinion here is in favor of hot water for private residences, and steam for large buildings, schools, public halls, etc.

A. Any heater that can take its supply at the top so that the flow of the water will be downward from the highest point will do.

There are cast-iron radiators extended with vertical fins made up of two, three, or four sections placed one above the other and connected at one end only. That is, the upper section connects on the right, say with the second section, and the second on the left with the third, and so on, giving these heaters the positive principle of a return-bend coil for hot-water work.

REMEDYING THE AIR-BINDING OF BOX-COILS.

Q. I REPLACED a cluster of old-pattern Gold's radiators with a box-coil, connected as in Figure 17. I cannot get all the pipes to heat, the centre ones being air-bound. What is the best plan to remedy the

FIGURE 17.

trouble? The steam and return-water all has to work to and from the boiler by the 2-inch main. The middle loops of each section of the box-coil do not heat. There is an air-cock in the lower header. The pipes which enter the upper header and its valves are one and a quarter inches in diameter, and the pipe from the lower header to the *tee* is one inch. Below the reducing-coupling the pipe is two inches.

A. Box-coils connected at top and bottom with a single pipe, as this one is, will always be air-bound, unless air-cocks are set all over it, or unless it leaks badly in the joints, the leaks acting as air-vents. Another difficulty about it is, that it can never be shut off entirely; in fact, we think the coil will work better if the valve shown is always kept closed and an air-vent put in the upper header To get satisfactory results cut out the nipple between the lower header and the *tee*, and put on a separate return-pipe with its valve.

HOW TO USE A STOVE AS A HOT-WATER HEATER.

Q. We wish to heat a long room by hot water if possible, heating the water by a large egg-stove. The idea is to run the hot

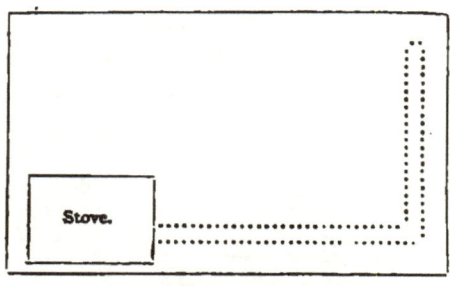

FIGURE 18.

water by pipes from the stove at one end of the room to the other end and return the pipes to the stove. The pipe will run on side walls of room. Will you be kind enough to say what pipe will be best, how to run it, and how to fit it up?

A. Use 1½-inch pipe or larger. Carry it from the coil, or water-back in stove, upward and as high as possible to an expansion-chamber, thence run downward and distribute around the room, about as shown in Figure 18. If possible, avoid carrying the end of the return-pipe where it goes to the stove upward for any considerable distance. If the pipe which first rises to the expansion-tank is covered so as to prevent much loss of heat it will help the circulation. The principle of the car-heater is what you require.

"PLANE" OR "PLAIN."

Q. WILL you settle the usage of the term *plain* or *plane* surface as applied to the outside surface of radiators?

In nearly all catalogues and other trade books it is spelled *plain*, but when used by writers it is generally *plane*. Why this difference?

A. The difference evidently comes from the view taken of what "plain" or "plane" surface is.

The trade was long accustomed to a certain kind of heaters or coils made of pipes or castings, which were smooth, or nearly so, on the outside; but then innovation came in and surfaces were put on the market which had projections on the outside, but too small to be "cored out" in the projection, and they were called "extended surfaces." The manufacturers of old styles of heating-surface then found it necessary to call their surface *plain*, to indicate that it was ordinary or common.

But soon again surfaces were made and put on the market which were not ordinary or common, but were corrugated or extended in such a manner that the inside or coring of the castings followed the contour of the outside of the radiators, and which could not be said to be "extended surface" in the first acceptation of the word. Nor were they the old kind, which had been called plain for the sake of distinction; but, nevertheless, they had the *distinction* which the advocates of the old school claimed as a point in favor of the pipe method—namely, that the inside or steam surface was the same, or nearly the same, as the outside surface, but as this latter was not *plain* it was called *plane*, to distinguish it from hilly or uneven surface.

Thus, pipe may be called *plain* or *plane* surface, while irregular shaped surface may be either *plane* or *extended*, according as the coring is done; but if the core does not follow the shape of the outside it is *extended;* therefore, all that is not extended may be called *plane*.

RELATIVE VALUE OF PIPE AND CAST-IRON HEATING-SURFACE.

Q. PLEASE state your opinion as to the relative value of pipe and cast-iron sections of the various manufacturers, for use as indirect surface in heating by steam.

In one instance which we know of there is an objection to the pipe-coils on account of their liability to become air-bound. The coils are about ten pipes wide by eight to ten deep, and three feet to five and one-half feet long, with headers at top and bottom for steam and return pipes. The coils have automatic air-valves, two to each coil. Large feed-pipes are the rule, and all well designed and put up.

A. There are no reliable data on the relative efficiency of commercial heating-surfaces. A box-coil should not become air-bound when it is supplied at the upper header with properly arranged air-valves.

If the only fault with the apparatus is air-binding, the difficulty is to be looked for in defective construction.

RELATIVE VALUE OF PIPE AND STEAM COILS.

Q. WILL you give an opinion as to the proper working of the coils shown in Figure 19? The following points will give you an idea of the construction of the apparatus as it has been put into a new building. The building has not as yet been plastered, and objections to the coils having been made, if they are to be changed it should be done now, while the building is in a rough state. Gold's indirect cast-iron surface was called for, but the steam-fitter, not finding the size of the brick chambers such as to fit the cast-iron surface well, made box-coils of inch pipe to take their place. If Gold's had been put in they would have had to be three to four sections deep, which would have brought the coil too near the water-line of the boiler.

The bottom of the pipe-coils is three feet one inch above the water-line. The coils are from three to six feet long, as in the sketch, from six to ten pipes deep, and from six to ten pipes wide (on the headers). For instance, take the largest coil, ten pipes wide, ten pipes high, and six feet long, or about 600 feet of pipe, or 200 feet of radiating-surface. Each coil has two air-valves connected at the end of the lower header next to where the return-pipe leaves it (see sketch). Do you consider these properly put up, so that there should be no trouble from air-binding in the middle of the coil, which loses the effect from part of the surface and renders the condensed steam liable to freeze?

What is the relative cost of the Gold surface compared with these coils of pipe as made?

A steam-pressure from two to ten pounds will be used, according to the season. There is plenty of pitch in the direction the steam travels to all of the pipes and coils, and the steam-supply mains are large. The air-valves are Davis's automatic valves.

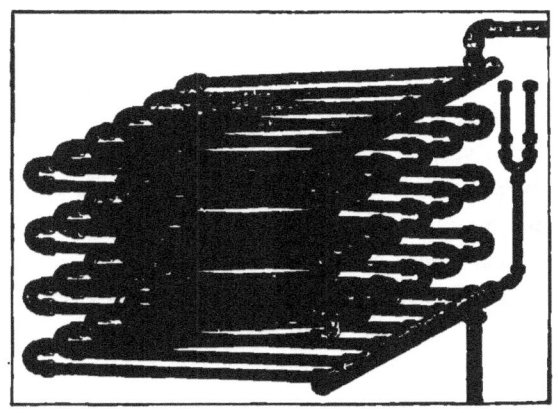

FIGURE 19.

Ought the coils to work free at one time and give trouble at another, or ought they work every day equally well?

A. It is difficult to reply to your letter directly. You do not say what the trouble is you found to exist, if any; you evidently only anticipate trouble. But to review the matter as far as we can from your letter, we will say, in the first place, that four sections of "pin-radiators," one above the other, can be placed in a height of about thirty-two inches. According to this, then, we cannot see why "pin-radiators" would not be further above the water-line than box-coils ten pipes deep, which must have about three and a half inches to centres, allowing for spread, which will make the height thirty-six inches.

As to which heater will work closer to the water-line—the pin sections or the 1-inch pipe box-coils—all other things being equal, the pin sections will give the best result, as the steam meets less resistance in them in passing from inlet to outlet; whereas the box-coil gives about the greatest resistance of any form of heater made, the steam having to travel through all its turns and long pipes of comparatively small diameter. Then, when a box-coil is used, it is better to have the steam and return at opposite ends of opposing headers, with the air-valves attached to either end of the lower header; but should the resistance and condensation in the coil reduce the pressure enough to allow the water in the returns to stand as high as the lower pipes of the coils, the air-valves will be abortive, no matter of whose make, and in such a case would be better attached to the upper header. Two air-valves, also, on one pipe are not likely to improve the working of the coil; one will accomplish all the two can if it is properly adjusted, but should the air-valves be separated, and one on each header or at opposite ends of the same lower header, some advantage may result.

There should be no pounding in the middle of the coils, and if it is found to exist it will not be caused by air-binding, but by the holding of water of condensation for the want of sufficient pressure above it to make it flow down regularly. It goes down spasmodically when the pressure of its own static head overcomes the resistance from below. Then it falls into hot steam in the lower header, condensation follows for a moment, a partial vacuum takes place, and the water-hammer or pounding is the result.

The liability to freeze is greater with common box-coils than with any kind of large-chambered heaters. The Gold surface is much cheaper than the plain box-coils, the cost being in the proportion of about four to five, and for this reason, if the full surface called for was used, economy or a disposition to save on the part of the contractor did not prompt the use of box-coils. With regard to pressures, box-coils work better at high than at low pressures.

Pitch alone to pipes will not make the water "go back." If a pipe runs nearly horizontal, with a pitch of half an inch in ten feet, and is then dropped suddenly one foot, it will work equally as well, for sizes such as are practicable in steam-fitting, as if it had the whole foot and more of pitch distributed through its entire length.

Coils should work every day equally well, but a box-coil that may work nicely at ten pounds by keeping the water down may refuse to work well at two pounds, and may let it up, simply because the amount of steam that can pass the supply-pipes and turns and pipes of the coils at the low pressure is all used to supply condensation, whereas with the higher pressures the greater velocity and increased density of steam for the same diameter flow-pipes will supply loss by condensation and friction, and have some energy left to "hold the water down"; in other words, the pressure from the boiler through the return-pipes will be nearly balanced by the steam-pressure through the coils, and the water in the pipes will rise comparatively little above its level at the water-line of the boiler.

WARMING CHURCHES.

Q. IN the matter of supplying *direct radiation* to churches, I would inquire what you think of placing a coil in each pew? Of course I do not lose sight of the necessity of the ventilating-shaft,

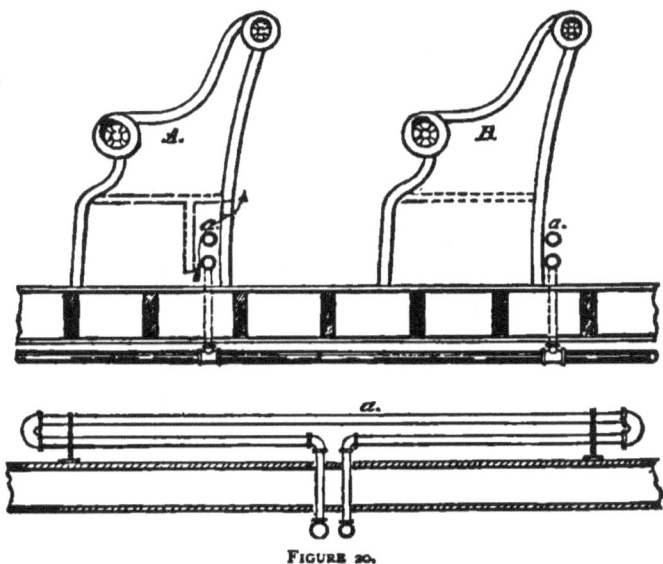

FIGURE 20.

or some other method of moving air, with a good proportion of indirect radiation for warming the supply of fresh air, but it does seem to me that a coil might be placed near the floor, and directly in the rear of each seat, so as to be at the feet of the people in the next seat behind, or under the seats, provided they are so constructed that the heat will not pass up in front of the seat under which the coil is, but to the pew *behind* the coil. Fearing I cannot make myself well enough understood, I send you the inclosed sketch (Figure 20).

These little coils, *a*, will connect with the steam and return main below the floor in nearly the usual way of large coils, without valves; but to avoid long return connections, I think the main return-pipe can be as well carried along side of the steam-main, as shown, than on the floor, as the basement is generally fitted for a class-room or chapel.

What do you think of the plan generally, and will not the warming of the room from below rather than from above be advantageous in this case, as it will allow each person to have a "foot warmer," so to say; and cannot a church be made more comfortable, even with a lower temperature, than by most other methods?

A shows one way, with a hanging partition, so the heat cannot pass front. B shows it with the coil in the pew, and for those who kneel, the stool can be made to cover it, if required.

A. We believe the First Baptist Church of Detroit is done somewhat in this way; at least, every pew has its coil. The plan is not new, though uncommon, and suggests several interesting questions, on which we would like the views of our readers.

The problem of church-warming is very unsettled. Churches can be made warm even to be disagreeable at parts, and be uncomfortably cold at the same time in other positions, and this happens with nearly all of our present methods, as no one rule can be made to suit all local conditions and styles of architecture. If pews run to the floor, the *counter current* which must always set in, sweeping to an outlet or a radiator with indirect radiation in the one case and direct radiators set at isolated positions along the walls in the other, will be interrupted, and will have to pass through the aisles, or on a line near the top of the pews—which should be the *neutral line*. On the other hand, when the pews *do not* run to the floor the return current is at the floor, and *over one's feet*. A person sitting in a pew which runs to the floor and is furnished with a door, with no heating source within it, with the pew in front of a window, will always be cold (in cold weather). The cold air will fall from the glass and settle in the pew, and will have to overflow to get out, leaving the pew always full of the air of the temperature at which it falls into the pew, and not of the temperature of the body of the church, the occupant being immersed to the middle in air much colder than the rest of the house. The same thing is experienced, though in a less degree, with the pews which abut on the piers between the windows.

WARMING CHURCHES.

Q. The plan of church-warming described in the letter preceding is not new nor uncommon in this vicinity. One church here has used that system for twenty-three years. I have also seen it used

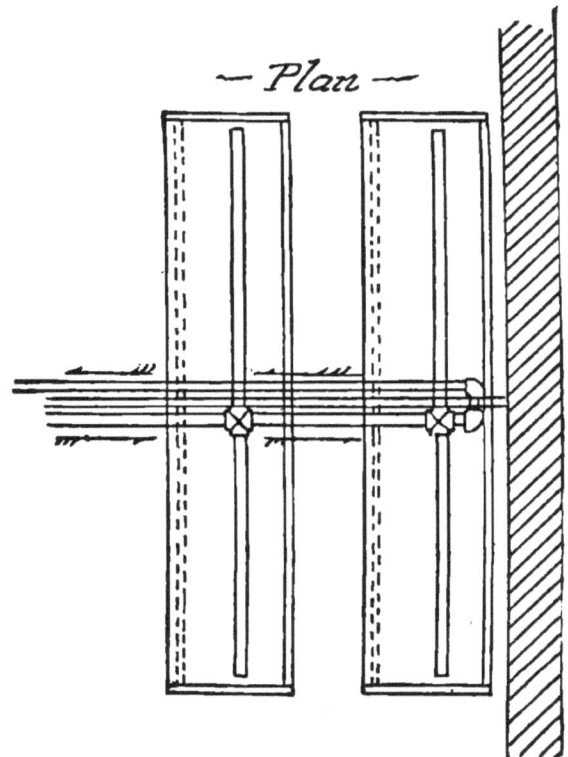

Figure 21.

in connection with small holes in the floor under the pipes connected with fresh air from the outside. A better plan, I think, is to run a single pipe on the floor to front end of church, either in the basement or on one side of the partition between seats (see sketch), thence back on the other side, with branches of 1-inch pipe for each seat, made like

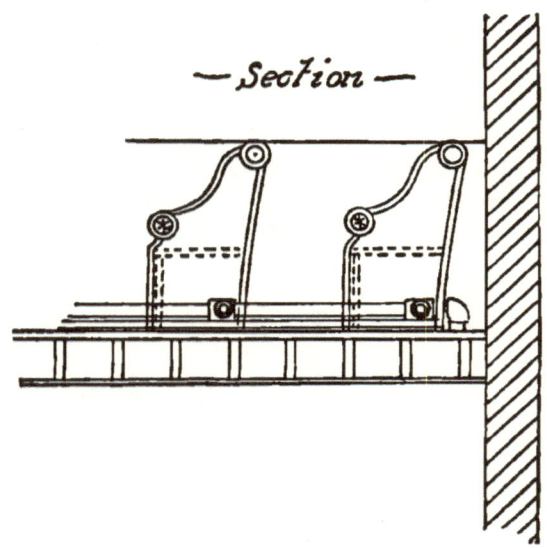

FIGURE 22.

a Nason tube. This partly overcomes the objection of having too much heat in the pews. I have another plan for warming and ventilating churches (where the basement is not used for class-rooms, etc.),

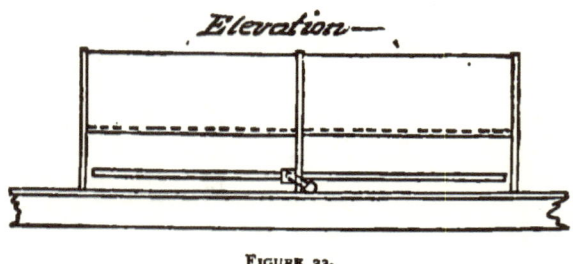

FIGURE 23.

that is superior to anything I have seen. When I can spare the time I will send you a description and sketch.

A. We think there is a little danger of having noise with this plan, unless the heating-pipe both ways from the *tee* pitched slightly upward.

FIGURE 24.

This is a matter of detail, of course, and all that is required is sufficient pitch toward the main pipe to secure the drain of the water that way, even should there be some slight change in the level of the floor—not an unusual thing with settling buildings.

PIPING AND FITTING.

STEAM-HEATING WORK, GOOD AND INDIFFERENT.

Q. I THINK it is evident that there is a good deal of steam-fitting done that does not represent the best practice, yet gives fair results.

Figure 25 is a sketch of part of a hotel job. You will observe that the steam-main is carried up from the boiler; also, that the radiator-

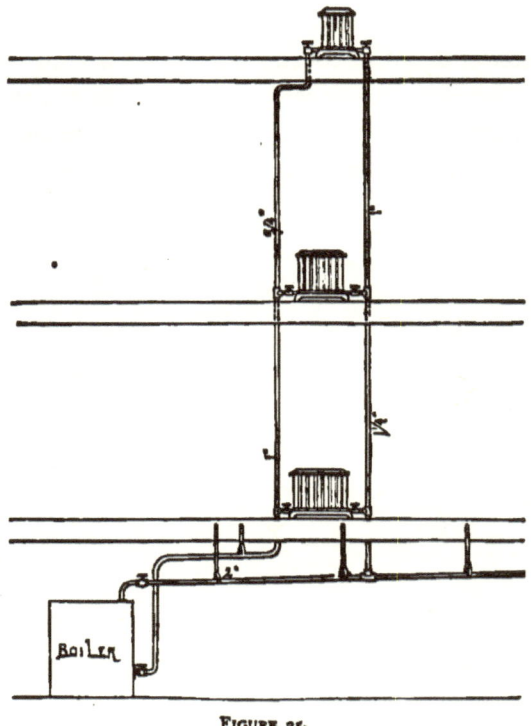

FIGURE 25.

connections are branched off straight from a T in the risers, making an almost rigid connection; also, that the returns do not come below the water-line, but are what are technically called "dry."

The steam-main is only 2-inch, yet there is at least 1,200 feet of heating-surface supplied by it.

The pressure carried is about five pounds ; the circulation is good —a little noisy at times, but that is the only fault.

Now, here is a piece of work with steam-mains too small, with radiator-connections that are contrary to all rules, with no reliefs, with no automatic air-vents, and with dry returns—in fact, contrary to everything that I have been taught to regard good work—and yet I can vouch for the fact that it heats the building comfortably, and that neither the man who did the work nor the man who had it done is aware that it is not a first-class job in every respect.

A. A very large share of all the steam-work through the country at large is done in a manner "just good enough to work."

Steam-fitting is like many other trades : its good points are only brought out by comparison. A man may appear comfortably and almost richly dressed in a suit of broadcloth made by a tailor who never had learned to *cut*, but nevertheless was a good *sewer*, and who had the hardihood to undertake any job that came along ; the cloth and trimmings being as good as other people's, *cost the same*, and if there was any saving it was on labor alone. Contrast him now with a man dressed by a man who *is a tailor*, and the comparison will be obvious and odious.

A similar difference exists between the work you show and the work done by a steam-fitter who is entitled to the name. At the same time we must bear in mind that heating-apparatus were invented to keep us warm, as were clothes, and that unless both are positively dangerous to the owner he should not be frightened about their appearance —*unless he is wealthy.*

PIPING ADJACENT BUILDINGS—PUMPS OR STEAM-TRAPS.

Q. I come to you once more with one of my sketches (Figure 26), which I hope you will be able to understand. I happened to see this job a few days ago, and remarked it could never be made to work. Last winter there was an Albany steam-trap over the boiler in the house No. 1, and it did all that was claimed for it, but the parties who own the house said that it required too much steam to run the trap, and the rooms in house No. 1 were made too hot in moderate weather; so they consulted a " jack-of-all-trades," who has advised the present plan. The bath-boiler (R), forty gallons, is intended to receive all the condense-water from houses Nos. 3, 4, and 5. Not less than 3,000 feet of pipe (1-inch) is used in the job. A Knowles pump is intended to pump the water back into the boiler.

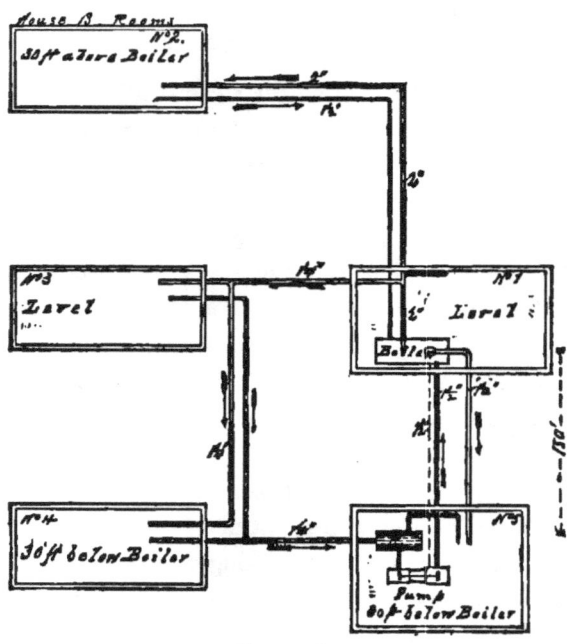

FIGURE 26.

There is no way to cool the water in the bath-boiler, and no way to get rid of the steam. Now, I wish to know if there is any pump that will pump with boiling water or steam. When the party started the pump it would not work, and he said it was because the pump was too small, and ordered a larger one. You will notice that houses Nos. 1 and 2 are on the return-system or low-pressure, while the other three will be on the high-pressure system, and I think very high-pressure, as the pump is thirty feet lower and 150 feet away from the main steam-boiler. There is no safety-valve on the bath-boiler. Is there not danger of this boiler blowing up? It is a cold place, and steam is not kept on it at night in the winter. There are all the necessary valves on the pipes, though they are not shown. The Knowles pump is No. oo, with ¾-inch suction and ⅜-inch steam and ½-inch discharge, increased at two feet from pump to 1½-inch.

A. If you mean when you say the Albany trap "did all that was claimed for it," that the apparatus heated and circulated properly, regardless of making the rooms too warm at times, we are of opinion that a great mistake was made when it was removed.

The graduation of the temperature of steam-radiators to suit changes of the weather is something that has been indifferently accomplished by the best engineers in the trade heretofore, and is not to be expected with such an apparatus as you describe. When an apparatus will work properly with any pressure, let it be one or forty pounds, then a good way is to carry low pressures in moderate weather, and higher pressures as it grows colder.

Speaking of the steam required to run the "Albany" or "Pratt" trap, or any gravity-trap, we think it is a mistake to use a pump instead which exhausts to atmosphere. Assuming the surface of the ball of such a trap to have seven square feet of condensing, it will condense less than *three* pounds of steam per hour, which is about one-tenth of a horse-power, and to do this it must be capable of putting about 720 pounds of water back into the boiler, regardless of pressure, whereas a pump to do the same requires the steam of nearly one horse-power, unless the exhaust-steam is turned into the heating-apparatus, which, of course, can only be done to advantage with low pressure in the pipes, when pump and trap are brought to about the same level, in point of duty.

A 40-gallon bath-boiler appears small for the purpose of a receiver for a pump, but should the pump be kept in constant action it will do, though if we designed a receiver for such a place we should proportion it to hold the condensed water that could be formed in an hour, to allow that much time for stoppage or an examination of the pump, etc. A galvanized-iron bath-boiler (if it is such that is used) will not burst with sixty pounds pressure, though, again, we must say we should not select it for a receiver. If the pump is sufficiently low to allow the hot water to flow into it by gravity from the bath-boiler (receiver), it should pump hot water. A pump is capable of forcing hot water, but will not "suck" it when it is near the atmospheric boiling point. The reason, presumably, the pump does not work well is, there is too great a loss of pressure in 150 feet of ½-inch pipe from the boiler. We would suggest a *one-inch* pipe for the pump.

We are opposed to giving an expression of opinion on the general methods of work—if they are not dangerous to life and limb—all the circumstances of which we cannot be acquainted with.

TRUE DIAMETERS AND WEIGHTS OF STANDARD PIPE.

Q. 1. Will you be kind enough to give the true inside and outside diameter of 1¼-inch steam-pipes, with the standard weight per lineal foot?

2. Will you also kindly inform me if it is possible for pipe to have the true diameter and thickness and not come to the standard in weight?

3. Which is usually the thickest, butt or lap welded pipe?

4. Will you also inform me if there is a pipe now in the market which is under the standard and which the unwary may buy and use as standard to the injury of his customers?

A. 1. The outside diameter of 1-inch standard steam or gas pipe should be 1.315 inches; its interior diameter is nominally 1-inch, but in reality a little greater, and its weight should be fully 1.67 of a pound per lineal foot. For 1¼-inch pipe, the outside diameter should be 1.66 inches, its inside diameter nominally 1¼ inches, but in reality a little greater, and its weight 2.258 pounds per lineal foot.

2. With butt-welded pipe—*drawn*—that is, not subjected to pressure in the manufacture, this might be possible to a very small extent, but lap-welded pipes—*rolled*—we think, should be fully up to the standard in weight if the thickness is maintained.

4. When the same thickness of "skelp" is used in both cases, we believe the lap-welded pipe is slightly thinner than wrought-iron.

There is 1-inch pipe now in the market which weighs less than 1½ pounds to the running foot, and when buyers do not know the makers from whom they buy, or when they buy through agents or jobbers, they should test and weigh their pipe so as to protect their customers and themselves.

EXPANSION OF PIPES OF VARIOUS METALS.

Q. PLEASE let me know at *what* point water-pipes expand.

A. Water-pipes contract or expand for every change in their temperature, let them be lead, iron, or brass.

Lead expands .0000158 of its length for each degree Fahrenheit it is warmed, and contracts the same amount for each degree it is cooled; wrought-iron from .0000066 to .0000074, according as it is soft or hard, and brass about .00001 for all qualities.

Example—Assume 100 feet of lead pipe warmed 100 degrees: we have .0000158 × 100 degrees = .00158 × 100 feet, = .158 of one foot, or $1\frac{9}{10}$ inches.

EXPANSION OF STEAM-PIPES.

Q. WILL you explain why it is that steam-pipes in buildings—rising lines—do not expand as much in practice as is given in text-books for the expansion of wrought-iron? For instance, I have an exhaust-pipe a little over 100 feet in height. The temperature of the building where it was put in was about 60° Fah., and now that an engine is exhausting through it, it has only elongated seven-eighths of an inch, instead of one and one-fifth inches, as I expected.

A. You have assumed that the pipe has been warmed from 60° Fah., or thereabouts, to 212°, and have calculated the expansion for a difference of temperature of 150°. This pipe being exposed to the temperature of the atmosphere on one side and the temperature of the steam on the other, will really have a temperature between the two. It is difficult to assume even how much the temperature of the iron of the pipe will be below the steam, or how much it will be greater than the air, though as a matter of fact it is nearly as great as the steam. It will vary also for different degrees of temperatures, and no reliable data that we know of are in existence on the subject.

When pipes are carefully covered with good non-conductors they are found to expand a little more than when uncovered.

The difference of expansion you wish to account for is but .32 of an inch, and this compared to 1.2 does not surprise us; though without what you say we should suppose it to be less.

ADVANTAGES CLAIMED FOR OVERHEAD PIPING.

Q. WE who do steam-fitting in New York City are seldom called upon to do work in cotton factories, but we know that in the Eastern States a practice has come into vogue of piping factories overhead—that is, the pipes are placed near the ceilings and near the outer walls—and it is actually claimed for the system that less pipe will make a building warm in this manner than if it were placed against the walls low down. How can this be so?

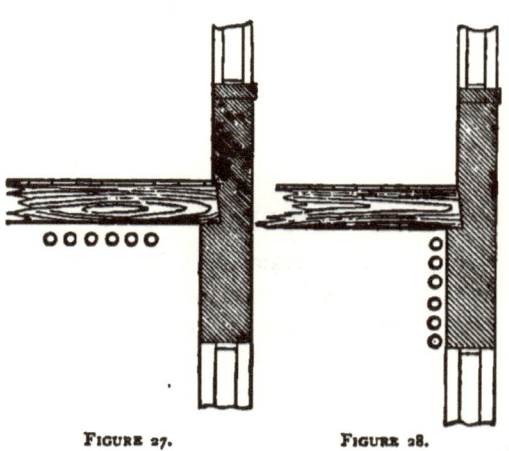

FIGURE 27.   FIGURE 28.

A. We believe two methods of piping buildings overhead have been been tried. One of them is to place the pipes in a nearly horizontal position, side by side, and a foot or two from the outside walls, and to run around the whole building, a section of which method is shown in Figure 27. The other, which is not so much used, is to put the pipes over the windows, but in other respects it is similar to the ordinary methods of placing coils under windows, and is shown in Figure 28.

In the case where the pipes are side by side (Figure 27), there is every reason to suppose more water will be condensed per foot of pipe than when the pipes are over each other (a full explanation of this is given in "Thermus" article, No. 20, page 458, Volume VII. of the *Sanitary Engineer*), and consequently more air is warmed in a given time, or a given quantity of air is made warmer in less time. With this method less pipe should do, but the amount saved, we should think, could only be determined by actual experiment. In the method shown in Figure 28 we think no pipe can be saved over the method of placing coils near the floor.

The whole system of warming overhead by direct radiation is more one of convenience and necessity than choice of position, but it has been found to have fewer objections than was first supposed to attend such a system.

POSITION OF VALVES ON STEAM-RISER CONNECTIONS.

Q. INCLOSED you will find a sketch (Figure 29) of the lower end of a steam-rising line. Why I trouble you with it is to point out an error that steam-fitters frequently fall into, and which is not always discovered until some damage is done by water, not to consider the noise that is produced if one attempts to shut off the line.

The error is this: The steam-fitter, in running his line, provides a *tee* between the steam-riser valve V and the flange-union for a radiator on the next floor, but when he comes to run his small pipes a and b he considers it is best to have the return-pipe b connected with the return-riser below the water-line, consequently he introduces a *tee* into the return, as shown, below the water-line and below the valve V'. What is the result? If you close the valves V, V', and V'' (as you must, if you want to shut off a rising line), unless the radiator is also *accidentally* shut off water from the return-riser r' below the valve V' will pass up in the pipe b through the radiator-base down the pipe a and into the steam-riser r. The result is the filling of the line with water, followed with a pounding noise. But this is not all; should one attempt to make repairs the water will flow out upstairs—apparently without reason, as the operator is positive he closed the riser-valves. Of course, when an engineer of a building finds it out, generally to his cost, he will ever after close the radiator-valves also, but I think you will agree with me in saying that this is not the proper way to shut off a riser, and that fitters should be more careful and put the *tee* at c; or if they want to have the first radiator take steam from the riser, they should put the return-valve V' at d.

A. We think the above letter fully covers the case; though if a *tee* were put into the return-riser in the nipple between the valve V' and

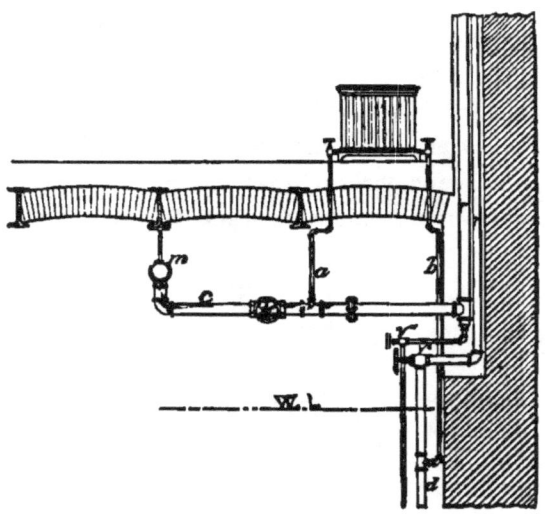

FIGURE 29.

the elbow, and b were connected with it, then the radiator would work equally as well as when the return is carried below the water-line for jobs of this class, as the condition is then only the same as exists on the next story above with a single return-riser, r', as shown in this case.

CAUSE OF NOISE IN STEAM-PIPES.

Q. (1) WHAT is the cause of the rattling and hammering noise in radiators and pipes for steam-heating on letting on the steam after it has been shut off over night or for several hours?

(2) Is it because distilled water has got trapped at some point?

(3) If so, would the admission of air at the lowest point of the system remedy it?

(4) It rights itself in about half an hour after steam has been turned on.

A. (1) The rattling and hammering noise (commonly known as the *water-hammer*) is caused by steam within the pipes coming in contact with water much colder than itself, producing condensation at the point of contact, and a vacuum more or less perfect, into which both the steam and water rush—the steam flowing with its accustomed high velocity, and the water, we will say, jumping as soon as inertia is overcome. The blow thus given when they meet is the cause of the noise, and exerts great strain and pressure on pipes of large diameters with much water and high pressures.

When steam can flow over water—as sometimes water is trapped in the base of a radiator—below the line of the inlet and outlet, a rattling only is experienced, with a rhythmic sound, caused by an imperfect contact of the steam and the water by having a stratum of air between them; but when the heater is full of water above the inlet, "banging" and "thumping" is caused by the water being forced asunder by the steam, but instantly returning with a shock to its original solidity on the condensation of the entering wedge of steam. This goes on until the water is sufficiently warmed to let enough steam pass through it to make a pressure sufficient within the radiator to press the water out at the return end.

(2) The distilled water, or water brought over mechanically, which cannot run away by gravity, will cause it.

(3) The admission of air will do no good in a poorly-constructed heating-apparatus.

(4) When the water is as hot as the steam, or very nearly so.

ONE-PIPE SYSTEM OF STEAM-HEATING.

Q. Does the one-pipe system of steam-heating have the preference in the East for blocks and large buildings? Here, where the thermometer reaches at times twenty-five degrees below zero, is this system better for circulation than the two pipes? The architects of this city specify the one-pipe system altogether.

A. The one-pipe system is entirely out of fashion, not only in the East but with nearly every one who has had a trial of it. The only advantage that can justly be claimed for it is cheapness of first cost, both in labor and materials. In small buildings (single houses) it may be used and give very little trouble, if well put in and used with very low pressures; but with high or medium pressures, or in large buildings, it will be an abominable nuisance. With all long runs of steam-pipes for heating purposes a circuit should be formed, and every radiator or coil in a building should "short-circuit" the larger one; the large or long circuit, or riser, being in turn a tributary circuit to the main circuit, or "mains." This keeps every circuit and subdivision of a circuit alive, and the interruption of any part of the system, such as the closing of radiators or rising lines, will have no effect on the remaining part of the circulation, and the opening and the filling of them again will be rapid, as the circulation they are a part of is active.

With single pipes there can be no circuit, except what goes on in the pipe itself. If the pipe is large and short this may do; but the limit is very soon reached, and air, water, etc., is impounded in the extremities. When the single valve on such a radiator is negligently closed or is defective, steam is admitted in small quantities and is condensed within the heater, partly or altogether filling it. If steam is wanted the valve is opened; but as the water is already in possession, and there is no way for it to run out but through the pipe at which steam tries to enter, there is naturally a conflict. Loud and continued noise is the result, and this noise will go on until the water in the radiator becomes nearly as hot as the steam, or until it can condense no more steam by contact. Then it will quietly run out at the bottom of the pipe while the steam is entering at the top, if the pipe is large enough for the two currents. Otherwise the noise is likely to continue at intervals, and what was intended for a steam-radiator will be a hot-water heater. Drawing a basinful of hot water from the air-cock and throwing it out of the window on a cold morning (the conventional method) relieves this condition of things for a short time.

Of course there are cases where these little air-cocks or valves are connected with a little pipe and run to the sewer or some other convenlent place where they "will not make a mess" and will be out of sight, and the occupant of the room is no more troubled with the fear that he or she may scald some one's child, as when they emptied their basin in the regular manner (the servant having refused to carry so much water down-stairs), and the inventor of the little pipe is blessed where he was before anathematized, and all goes apparently well. But, from a practical point of view, what has this man done? He has altered a single-pipe system into a two-pipe system, and a very bad one at that, as he is losing a large part of his condensed water into the sewer, or, if it does not go to the sewer, he is sapping the foundations of the house with it. The result is, that if he has not a good water-feeder, there is a burned boiler in a short time, and no one knows where the water went to, and if the steam-fitter has an idea he is likely to keep it to himself.

It is also difficult to get air from a one-pipe system.

HOW TO HEAT SEVERAL ADJACENT BUILDINGS WITH A SINGLE APPARATUS.

Q. I HAVE made a study of the systems of steam-heating for the past year, and have been rewarded with the most gratifying results, but just now I have a problem which seems difficult to solve. It is as follows:

A gentleman desires to have a factory and private residence heated by direct vertical radiators from one source—viz., the factory. The factory is elevated about four feet above residence, as you will see by Figure 30, and they are ninety-four feet apart. The factory cellar is eight feet deep, and if a line were drawn from the top of the residence cellar it would come within four feet of the bottom of the factory cellar. What I wish to know is, whether the residence can be successfully heated by a gravity system of steam-pipes, provided I could place the water-line of boiler one to two feet below steam-main. Would this be practical? I suppose I could overcome the difficulty by the use of an automatic return-trap. Any information you can give me will be most gratefully received.

P. S.—Would a gravity system in the factory and an automatic return for the residence work successfully together?

A. One or even two feet between the level of the mains of a gravity steam-apparatus seldom gives satisfaction. Not that it is impossible to have the apparatus work if you use very large main pipes, but there will be times when comparatively sudden changes of pressure will raise the water-columns in the return-pipes, and then it would be very difficult to shut off or let on steam without considerable noise and with the chance of breaking fittings or pipes by the water-hammer when the water-line is so close.

We know of one building—a large apartment-house—in New York working on 20-inch difference of level between the second gauge-cock and the mains. The apparatus works well when untouched, and steam is raised and lowered with all valves opened, but should steam be gotten up on the boilers with the main valves closed, or should the valve be closed for some purpose for a short time, it is not then safe to let steam on the building again unless all the water is run from the return-pipes into the sewer. Again, should a riser be shut off and let on again with an apparatus that is run so close, the sudden draught of steam into the empty riser and heaters will cause a momentary loss of pressure in the mains sufficient to let the water-columns up.

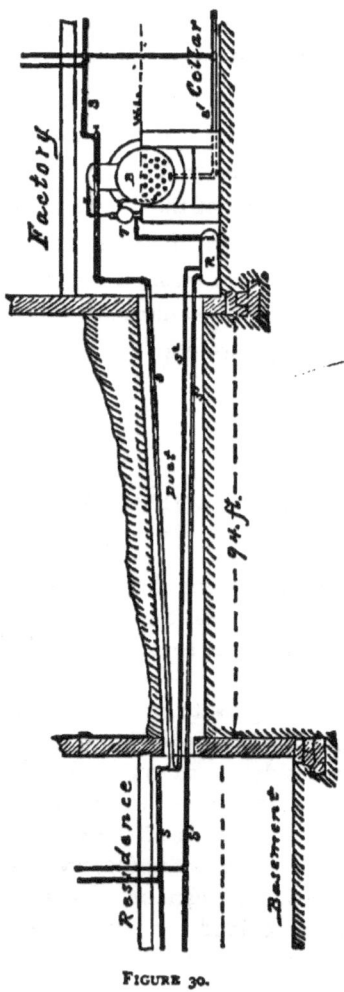

FIGURE 30.

In this case, all things considered, we would advise the use of a gravity system in the factory, as shown to the right in the diagram, and a direct return-trap system in the residence.

The pipes S are the steam supply-pipes or mains, and S^1 indicates the return-pipes, while S^2 is the relief from the end of the main underground back to the receiver of the trap. This pipe when it enters the receiver should be provided with a check-valve, as well as the pipe S^1, to prevent "short-circuiting" or back-pressures from the receiver to the mains, before the pressure from the boiler filled them equally when letting on steam.

The residence may be piped with the return-pipe overhead, as shown in the diagram, but the sizes of pipe used should be very nearly as great as would be used in a gravity apparatus of the same size.

PATENTS ON THE MILLS SYSTEM.

Q. PLEASE be kind enough to let me know if there is a patent on what is called the "Mills system of steam-heating"—*i. e.*, running the steam-main to the top of the building and running the distributing-mains downward?

A. There is a patent or patents on the "Mills system of steam-heating," but the Mills system patents, as we understand them, do not cover the right to use steam fed through a down system when in connection with a separate return riser-pipe. The Mills system is the use of a "down-steam" riser, with a short connection with one valve to one end of each radiator, the return-water flowing through the same connection into the riser again, and falling through it into a horizontal return-pipe near the floor of basement or cellar.

AIR-BINDING IN RETURN STEAM-PIPES.

A CORRESPONDENT writes:

"SIR: It is generally known to the steam-heating trade throughout the country that air-binding is likely to take place in return-pipes of steam-heating apparatus and be the cause of continual annoyance by the holding of the return water in vertical rising lines or radiator-connections much more above the water-line than is due to the difference of pressure between the boiler and the ends of the distributing pipes of the steam-supply system. I have run return-pipes along basement-floors in gravity apparatus sometimes, with a *pitch* in the direction of the boilers when I can conveniently do so. Often I am forced to run level, and now and then I am forced to rise up and go down again to get over something that is in the way. This last, I believe, many steam-fitters do as well as myself, claiming that it can have no perceptible effect on the working of the apparatus. In one case, at least, I have discovered it to be a serious matter.

"The apparatus in question I had fitted up with a sufficient grade to the return-pipe, until I had passed half the length of the basement, when I was forced to rise above a drain-pipe or go under it. I chose the former method, giving the matter very little consideration at the time. When the apparatus was completed, and steam up, I was surprised to find that the water stood much higher in certain return-pipes than I had calculated on, and that at times water was running from the air-valves on *certain* rising lines—the valves being placed at the lower end of the lines near the ceiling of the basement. These rising lines and radiator-connections which acted in this manner were the ones furthest from the boiler, and at once it occurred to me that I had by some means used mains of two small a diameter; hence the supposed loss of pressure. But on mature deliberation I assumed my mains were large enough, and I began to look for my trouble elsewhere. After proving there was no mechanical stoppage in the pipe, I filled up and tried my apparatus again, and found it went well for a time, but again filled up some distance in the same risers.

"The question then came to my mind whether 'air-binding,' such as you sometimes have in waste and water pipes, could have anything to do with it, as I had noticed that the rising lines between the boiler and the rise in the return worked well, though the others did not. To test the matter I punched a small hole in the return-pipe, when compressed air immediately rushed out and the water came to its proper level in the other rising lines.

"This experience may be of service to some of your readers who are troubled with imperfect circulations in their heating apparatus, and I offer the suggestion for their benefit."

AIR-BINDING IN RETURN STEAM-PIPES.

A correspondent writes:

"Sir: Referring to the preceding letter on this subject, it is usually advisable to run the return-pipe *under* such an obstruction as your correspondent mentions, by constructing an inverted syphon in the return-pipe, as in Figure 31. If the return-pipe cannot be passed under the obstruction, it must be passed over it,

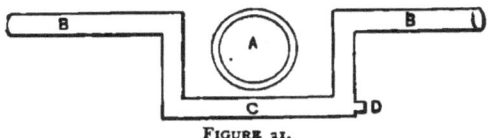

FIGURE 31.

A—Obstruction. B—Return-Pipe. C—Inverted Syphon. D—Plug or Cock for drawing syphon when necessary.

as in Figure 32. Your correspondent omitted to put in an equalizing-pipe connecting the top of his syphon with the nearest available steam-supply pipe; consequently when the water was run into the apparatus, the lower parts of the return-pipe filled, and forced the air into the top of the syphon, where, as it had no outlet, the "air-binding" was the

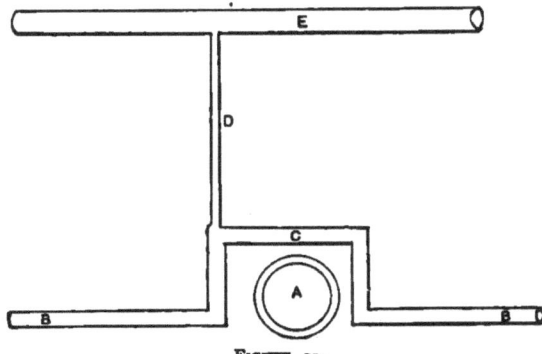

FIGURE 32.

A—Obstruction. B—Return-Pipe. C—Syphon. D Equalizing-Pipe. E Steam-Supply Pipe.

result. If your correspondent will put in an equalizing-pipe, as in Figure 32, he will find the air-binding permanently cured. When pipes are filled with water, the air will always lodge at the highest point if no outlet is provided for it."

VENTILATION.

SIZES OF REGISTERS TO HEAT CERTAIN ROOMS.

Q. Would you be kind enough to give me the sizes required for the following registers? I am putting in registers too large in a house, where they will look clumsy. This house is in Salem. It has an orchard on the north, with a large brick stable to keep off cold winds. For chamber marked D. R. Chamber he has 17′x25′ between the opening. I say it is unnecessary. Also, how many square feet of radiation would your sizes require, and how often would this change the air in the room? The walls of the house are hollow.

Size of register which is required to heat the following rooms by indirect radiation, boiler 3′x11′, thirty 3-inch tubes, on north end of house:

		Cubic Contents of Room.	Square Feet of Glass.	
N.	Dining-Room.	5,661	150	Over Boiler.
E.	Back Parlor.	4,736	90	Next to D R opens into it.
S. E.	Front Parlor.	4,390	69	
S. W.	Library.	3,172	75	
W.	Den.	2,181	48	
Second Floor.				
N. E.	D. R. Chamber.	2,747	112	
N. W.	Kitchen Ch.	3,124	115	
W.	Den Ch.	1,896	23	
S. W.	Lib. Ch.	2,520	64	
S. E.	F. Parlor Ch.	3,591	60	
S. E.	Dressing Room.	1,071	23	

All of these rooms have open fire-places except the dressing-room.

A. Under conditions such as you are likely to have in the rooms on the first floor of a house, with indirect radiation and natural currents, with a fire-place chimney, the velocity of the air through the flues will be between 1½ and 3 feet per second. If you now consider a flue of one square foot of cross-section you will have 3,600 cubic feet of air that will pass into the room in an hour, assuming your velocity to be only one foot per second; but assuming it to be 1½ feet, the minimum, you will have 5,400 cubic feet passing, the equivalent, very nearly, of moving the air once in an hour in the dining-room. If four healthy persons occupied this room continuously, this would give them *fair* ventilation, and should you get the velocity of three feet per second, it would give them *good* ventilation.

If this amount of air entered at the temperature suitable for living and breathing, say 65° or 70°, and the temperature outside was 10°, the loss of heat through walls and windows would be such as to keep the room at a temperature much too cold to live in. To make the room, therefore, fit to live in (air at 70° or thereabouts), air must enter in very much larger quantities, or it must enter at a temperature much above 70°, and you must trust to mixing it with the air cooled by the windows and walls to maintain a living temperature. If we assume now that each square foot of glass will cool 1½ cubic feet of air per minute from the inside temperature to the outside temperature, we have for dining-room 150 square feet glass × 1½ = 225 × 60 = 13,500, of the number of cubic feet of air cooled, say, from 70° to 10°, or by 60° Fah.

This gives us 13,500 cubic feet of air warmed (or cooled) 60°, or 810,000 cubic feet warmed or cooled 1°, as the amount necessary to maintain the heat. Of this amount we require at least 5,400 cubic feet of air warmed from 10° to 70° to maintain ventilation, and as only 5,400 will actually come through the register or flue with a velocity of 1½ feet per second, we must admit it warmer, and this will give us 13,500 × 60 = 810,000 ÷ 5,400 = 150° as the temperature at which the air should pass the register for such conditions. But 150° Fah. is a temperature that cannot be readily obtained from ordinary steam-coils, though it can be from a furnace, and 100° to 120° is all that can be looked for with ordinary steam-apparatus.

As we are now forced to take air at 100°, we have 100° ÷ 810,000 = 8,100 as the number of cubic feet of air at the temperature necessary to maintain the heat of the room, and the heat we must maintain as well as ventilation. This divided by 5,400—the quantity of air that will pass safely through one square foot of flue—gives 1.5 square foot as the cross-section of the flue for the dining-room. This, fortunately for us, gives air in excess of *fair* ventilation, and what may be called almost *good* ventilation.

We are required now either to get a greater than a minimum velocity of air through a given size flue, or to provide a flue of greater size, for we are forced to take air between 100° and 120°.

Therefore, with a minimum velocity of 1½ feet for first floor and three feet for second floors, we have approximately flues of one square foot of cross-section for each 2,500 cubic feet of space in first-story rooms, and the same for 5,000 cubic feet of space in second-story rooms and those above, when we allow for a practical magnitude to overcome friction of turns, etc. No flue should be less than 8x12 inches.

To find the coil for a given room or flue, take all the air passed for an hour—say 8,100 cubic feet for dining-room; multiply it by the degrees it is warmed, and divide it by 48,000, and it gives the number of pounds of water to be condensed in the coil in an hour. Average coils and radiators will condense from one-fourth to one-third of a pound of water per hour per square foot of surface.

Let the open fret-work of the register have equal area with the flue.

On page 32 of Tuttle & Bailey's catalogue of registers will be found the capacity in square inches of openings through fret-work. With forced ventilation, flues may be very much smaller.

DETERMINING THE SIZE OF HOT-AIR FLUES.

Q. CAN you give any rule for determining the size of a hot-air flue with reference to the cubical capacity of a room? To illustrate: say on first floor, one square inch of radiating surface to one cubic foot of space, three-quarters for second floor, and less for upper floors. Now, on this basis, what is the rule for determining size of the ducts and cold-air inlet, and proportions between them?

A. To aid in the computation of dimensions of flues the following was published in the Pascal Iron-Works Catalogue in 1870 :

"The following dimensions of flues will insure a supply of warm air :

"*For the heating-flues.*—Height of bottom of register above upper surface of radiator, 1, 2, 3, 4, 6, 8, 10, 15, 20, 25, 30, 40 feet and above ; square inches of flue needed for each square foot of radiating surface which the room requires, 2, 1.41, 1.16, 1.00, 0.82, 0.71, 0.63, 0.52, 0.45, 0.40, 0.35, 0.32 inch.

"To the area of cross-section obtained from these figures add twenty square inches to compensate for resistance of mouth of inlet and of discharge, or to give practical magnitude to small flues.

"*For the ventilating-flues.*—The same rule may be followed, only the height is to be taken from top of register to top of chimney or ventilating-stack or outlet."

Heating-flues should be tin-lined.

Example—Room of 3,150 cubic feet capacity, latitude of Philadelphia, north-west exposure, first story. Ratio of one square foot radiating surface to 63 cubic feet space, where the glass window-surface in the room is not over one of surface to 100 of space, would require 50 square feet of radiating surface for steam not over 15 pounds pressure. Suppose top of radiators to be two feet below bottom of registers (or surface, if they are flat) $\therefore$ $50 \times 1.41 = 70.5 + 20 = 90.5$ square inches, or a heating-flue 9 x 10 inches would be demanded. Suppose top of ventilating-chimney to be 40 feet above top of ventilating-register $\therefore$ $50 \times 0.32 = 16 + 20 = 36$ square inches, or a ventilating-flue 9 x 4 inches would be needed.

These figures are wholly empirical, and the 36 square inches is evidently too small for an outlet, when the inlet to the room is 90.5 ; but they will serve to guide a practical man who has had experience in heating in proportioning his requirements on the builder.

Cold-air ducts supplying air to numerous hot-air registers can safely have a cross-section as large as the sum of all the heating-flues, but it will be found that the inducement of the high flues will allow the cold main or duct to be *throttled* by some kind of shut-off with advantage, while in any steam or hot-water apparatus some automatic contrivance should close the cold duct whenever the steam or heat of the water goes down.

WINDOW-VENTILATORS.

Q. ONE of the teachers of an institution for girls has applied to me to know whether any improvements have recently been introduced in methods of continuous window-ventilation. The method by introducing into the sash a revolving fan is of course familiar. Some one patented, a few years ago, a device for "filtering" the air which is permitted to pass through wire gauze at the base of a window, allowing a constant, gentle draught, free from dust, or by substituting muslin, free from moisture, to some extent if the muslin be changed from time to time.

This latter method was introduced into one or two schools of New York City by joint recommendation of my friend, Dr. W. Gill Wylie (40 West Fortieth Street), and myself. Dr. Wylie suggested to me to write to you on the subject. If you can kindly refer me to any articles containing an account of further improvements, I shall esteem it a favor.

A. There are many devices for securing ventilation by windows without producing unpleasant draughts, the principle of all being the same—viz., to direct the incoming current of air upward toward the ceiling by means of a deflecting-plate. Wire gauze and coarse muslin are used in many of these contrivances, to keep out dust, flies, etc., and to break up the incoming air into fine streams, and thus avoid draughts.

The latest patents of this kind we have seen are that of J. G. Bronson (No. 270,733, dated January 16, 1883), for an extra sash or deflector-plate, and that of Sarah B. Stearns (No. 271,146, dated January 23, 1883), for a deflecting-plate with a gauze or cloth strainer. Copies of the specifications and drawings for these patents can readily be obtained from the Patent Office in Washington.

The practical working of such contrivances depends on the mode in which the room is heated, on the presence of an open fire-place or special foul-air flues, on the external temperature, and on the direction and force of the wind. As adjuncts to a properly arranged system of ventilation they are convenient and useful, but it is a mistake to rely upon them solely to secure ventilation in a room containing a number of persons, as for example, in a school-room.

Some additional particulars about simple means of window-ventilation are given in the abstracts of Dr. Lincoln's paper on school-houses in the *Sanitary Engineer*, Vol. VI., page 186, and by Dr. J. S. Billings, Vol. IV., page 130.

WINDOW-VENTILATORS.

A CORRESPONDENT writes :

"I notice an inquiry regarding window-ventilators. Having had occasion to use something of the kind, I have devised and used with very good results a modification of the old plan of setting a strip under the sash, for the sake of the air entering at meeting-rails, of which please find a sketch (Figure 33) inclosed.

FIGURE 33.

"It consists simply of a piece of board as long as the width of window and three or four inches high, set about an inch back of the sash and secured in place by vertical grooves in the ends, sliding over two round-headed screws in the stop-head at each side of window.

"By raising the sash a trifle, air is admitted in a thin sheet and deflected upward, and there is the same action at the meeting-rails, which is all I have ever seen attempted by anything of the sort. A wire-gauze screen can be bracketed to the outside of the strip if desired.

"The cost is nominal; there is no interference with using and fastening the sash in the usual way; the strip can be removed in an instant, and whether it is in or out, the window is not disfigured."

REMOVING VAPOR FROM DYE-HOUSE.

Q. COULD you inform me of the best way to take steam out of a dye-house? It is easy enough to do this in the summer by opening the windows and letting it blow out, but in the winter the cold air prevents it from coming out.

It has been tried to draw it out with a fan, but that does not answer the purpose. The fan, of course, draws out the air, but leaves the vapor in the house. We have also tried to ventilate the steam through the roof by artificial heat, but with the same result as the fan.

This is a dye-house for a hat factory, and the steam is of a very wet nature, consequently like a vapor, and not like the ordinary steam from a boiler.

A. We think a properly arranged fan would do more to remove the vapor from the room than any other means, excepting, perhaps, an aspirating-shaft

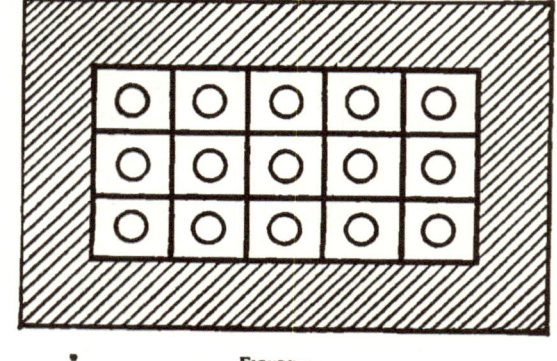

FIGURE 34.

of large dimensions and high, with heating surfaces within the shaft, arranged with a view to getting the greatest results from the smallest quantity of heating surface.

In any case provision must be made for admitting as much air, and as fast as the fan or aspirator is capable of drawing it out.

The air is the vehicle which holds the steam or vapor in suspension, and when it becomes surcharged, the vapor being the lightest, the superabundance will be found near the ceiling, so that whether a shaft or fan be used, the air must be drawn from the top.

A way to remove moisture from the air of a drying-room, without changing the air, is to condense it against pipes, placed at the upper part of the room, through which cold water is circulated.

Troughs are arranged under these pipes to receive the water so formed and conduct it without the house.

Professor William P. Trowbridge, of Columbia College, in a paper read at the first regular meeting of 1882 of the American Society of Mechanical Engineers, says: "There seems to be no doubt that steam-coils properly devised and adapted to chimneys or flues will give a more efficient ventilation than the blower, for less cost of construction and maintenance;" He also says: "The arrangement of the steam-pipes in such a manner that the greatest amount of heat will be transferred to the air with the least resistance to its motion is a matter of importance;" and he suggests that a flue may be divided into smaller flues at its base with sheet-iron diaphragms, between which the vertical pipes should be placed, as shown in the diagram, Figure 34.

VENTILATION OF THE CUNARD STEAMER "UMBRIA."

The steamer "Umbria," the latest addition to the fleet of the well-known Cunard Company, reached New York on her maiden trip on November 10, 1884. She was built at the works of John Elder & Co., near Glasgow, Scotland, and is 520 feet long by 57 feet 3 inches beam, with a depth of hold to upper deck of 41 feet; her measurement being over 8,000 tons, and, though not the longest, is probably the largest vessel afloat, except the "Great Eastern." She is built of steel, and in the finest manner known to marine architects. She is divided into ten water-tight compartments, with doors sliding across the ship instead of moving like a portcullis, or swinging.

Her engines are of the inverted compound type, being composed of one high-pressure cylinder, 72 inches in diameter, between two low-pressure cylinders, each of 105 inches diameter, one fore and the other aft of the primary cylinder—the stroke of all being six feet. The piston-rods and connecting-rods are of forged steel, the former being 11¼ inches in diameter and the latter 15 inches. The main shaft is 24 inches in diameter, and has a thrust-bearing of 17 feet in length, with as many collars on the shaft, which run in bearing-shoes, each of which is capable of separate adjustment.

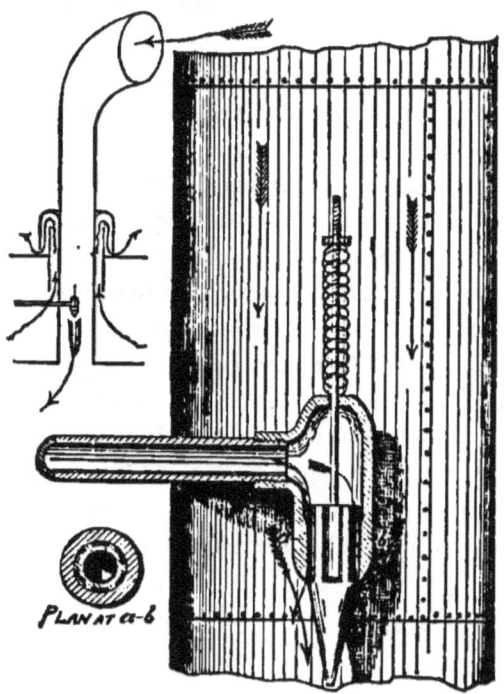

FIGURES 35 AND 36.

She has accommodations for about 800 first-cabin passengers. Her music-saloon is 76 feet long by the full width of the ship, and is about 8 feet in the clear underneath the deck-beams. The dining-saloon is immediately below the music-saloon, and is of the same dimensions; a domed skylight giving downward light to both through a well-hole of equal dimensions through the deck of the saloon.

Systematic ventilation and artificial warming are provided to all parts of the ship, the sailors' forecastle, even, and the firemen's quarters being provided with both. The system of ventilation used is known as

Green's patent. In the ordinary ship's ventilators are placed injecting nozzles, Figures 35 and 36, through which air, at a pressure of five pounds per square inch, is discharged, causing an induced current of air either in or out of the compartments of the ship, or both, as the case may require. An air-compressing engine is provided, and situated in a part of the main engine-room set apart for it. It is supplied with steam from the main boilers, and also from a donkey-boiler to be used when in port. This engine compresses air within reservoirs, from which it is released into the injecting nozzles through pipes of about one inch internal diameter, which lead from a trunk-main which runs the whole length of the ship on both sides. Outward movements of air are secured through the hollow steel masts and through the annular spaces between the smoke-stack and jackets which surround them.

Each stateroom is not supplied with a separate ventilator, but at short intervals along the passages they are to be found.

The heating is done by copper pipes of about three inches in diameter, over which is placed a fret-work guard of cast brass. These heaters are principally in the passages where the cool air is admitted, and heat and a change of fresh air to the staterooms is secured by means of fret-work between the transoms and fixed lattice-work panels in the lower parts of the stateroom doors, causing the heat and air to enter by the latter and escape by the former.

Figures 37 and 38 are details of the heating-pipes and guards. They are the result of circumstances, and are efficient; but an improvement that we would suggest would be to have them so arranged that a steward, in cleaning, could remove the guards to properly cleanse under them—which might be done by having them hinged at one side, that they could be turned over. As it is, where a hose can be used, they can be kept ordinarily clean; but in the stateroom passages, and under settees and dining-room tables, this cannot be done, and the scrubbing or sweeping which has to be resorted to forces dust, etc., deeper into them. Steam is taken from the main or donkey boilers, as the case may be, and the condensation is returned to the hot well or used for washing purposes aboard ship.

For the purpose of heating water for the baths and for washing when there is not sufficient condensation from the heating-apparatus, or when it is not in use, two special condensers are supplied, and a still also is provided for making drinking-water, should the supply that is carried in tanks be not enough, as on a long or delayed voyage. Two of Haslen's (of Leeds, England) dry-air refrigerating-apparatuses are provided—one small for the storerooms, and one large for a cargo compartment. Their principle is to compress air into reservoirs, through which the sea-water is circulated in small pipes—somewhat in appearance like a surface condenser—for the purpose of extracting the

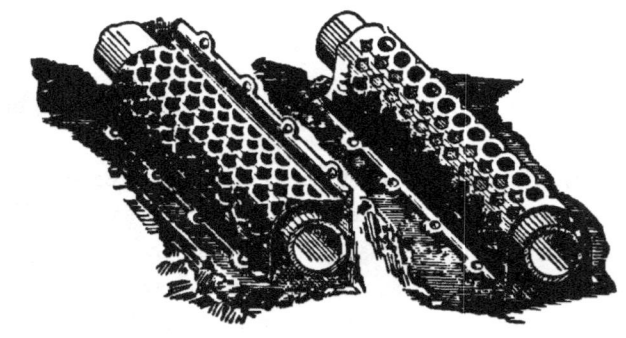

FIGURE 37.   FIGURE 38.

heat caused by compression. The air is then liberated, when it immediately expands, and is in condition to seize on the heat of surrounding objects, even to the point of producing ice. The actual capacity of the large machine is not known to the engineer, but, as an experiment, 130 tons of dressed meat were taken to Liverpool on the return voyage, preserved this way.

The ship is lighted throughout by incandescent electric-lights, the system being Andrews', of Glasgow. The dynamos are Siemens', and are four in number, each driven by one of Brotherhood's three-cylinder engines.

The vessel is commanded by Captain Cook, and her chief engineer is Mr. John Heggie.

CALCULATING SIZES OF FLUES AND REGISTERS.

Q. CAN you tell me if there is any published data giving the size of heating flues and registers for steam and hot-air heating?

A. The size of a flue in a wall or the opening through a floor will all depend on the *amount of air required in a room in a given time and the velocity* you are likely to obtain with any particular apparatus. The flues or registers for hot-water apparatus should represent the maximum of size, as the temperature of the air-currents will be lower than with any other class of heating-apparatus, and may be said to represent the minimum of temperatures. On the other hand, the air from a furnace being warmer than from a steam or hot-water apparatus, the minimum of flues may be used. This, of course, is all on the assumption of natural currents—no forcing, as with a fan, being used.

The force which produces motion in a heating-flue is the difference between the weight of a column of warm air in the flue from its start at the heating-coil or furnace, until it reaches the outside air at the top of the house, and a corresponding column of the outside atmosphere of whatever temperature it may be. If the inside column of air be twenty-five feet high and you warm it 120° Fah., you increase its bulk one-quarter—or, in other words, it will have to be 31¼ feet high to give it equal weight with a column of air of the density from which it had been warmed. But as the height of the heated column is limited by the height of the flue—twenty-five feet—the force of the cold column presses in on it with a velocity equal to that acquired by a body falling 6¼ feet; the velocity of the descent equaling *eight* times the square root of the height of the descent in feet or decimals of a foot, or $\sqrt{g \cdot h}$ 16.09 × 6.25 = 20 feet per second as the velocity in the flue; in which g is the distance through which a body falls in a second of time and h the distance fallen through. Presumably in practice one-half this velocity cannot be exceeded, and some authorities claim a coefficiency of .4 as about right for ordinary circumstances of flues and registers.

For more information on this subject see Billings' " Ventilation and Heating," page 31, and Hood's " Warming and Ventilating," page 359.

CHURCH VENTILATION.

Q. WHAT would be the best mode of drawing the hot-air from between two roofs of a church building, the space being between ceiling and roof?

If a screw is recommended, what would be the circumference or diameter?

I would like to attach a windmill to draw the hot-air out.

A. If it is simply to cool the space between the ceiling and the roof of a Gothic structure, we think louvered windows at the ends and a louvered ventilator at the apex should suffice.

If it is necessary that the vent-flues of the church or the ventilators over the chandeliers open into this space, and there is no artificial outlet, make suitable outlets similar to the above.

If, again, you wish to remove warmed or vitiated air from the church faster than it can go out by such flues or openings as you may chance to have, by natural currents, a fan may be used to good advantage.

The size of the fan will depend on the amount of air to be changed in a given time, and a fan of the class used in the Capitol at Washington, five feet in diameter, will move from seven to twelve thousand cubic feet of air per minute according to the speed at which it is run.

There are other classes of fans in the market, of which the makers will be glad to furnish the capacity and size, if you state the quantity of air you wish to remove.

STEAM.

ECONOMY OF USING EXHAUST STEAM FOR HEATING.

Q. I have in charge two tubular boilers 4½'x14', 70 pounds steam pressure, and an engine 22"x 36", adjustable cut-off one-half, 78 revolutions. The exhaust steam from the engine heats three floors 60'x 40'. I am about to connect a Korting condenser to the engine, and get about 28-inch vacuum; this will relieve the engine of all back-pressure, besides a gain of power. Then I will heat the three floors by live steam, and return the same directly to boilers by means of a Pratt return-trap. I will, also, give them more heat than by using the exhaust steam. I will use a pressure-regulator, and give seven pounds to heat the building. I would like some information as to what the economy will be in favor of using the condenser and trap over heating by exhaust steam.

A. To avoid a misunderstanding about the question of using exhaust steam for warming, we will say that when all, or even a comparatively small quantity of it, can be utilized and properly condensed, economy is in favor of throwing a slight back-pressure on the engine and using the exhaust steam. The commoner and poorer the engine the larger the ratio will be in favor of utilizing the exhaust steam for heating.

In the present case, cutting off at one-half stroke, the mean pressure in your cylinder may be assumed at 57 pounds per square inch when there is one pound back-pressure above atmosphere on the engine, which will develop a *total* (not indicated) horse-power of 304. If, on the other hand, you expand your steam down to 14 pounds below atmosphere, your mean pressure will be seventy pounds in the cylinder, with a total horse-power of 374, which will be a gain of 70 horse-power, without evaporating any more water. If we assume you *now* use 45 pounds of water per nominal horse-power, your total evaporation will be 13,680 pounds of water, whereas if you develop 374 horse-power under the same conditions (non-condensing), you would have to evaporate 16,830 pounds of water—3,150 pounds more. We thus appear to gain 23.5 per cent., but when we consider that we lose 11.5 per cent. due to the difference of temperature of feed-water between 212° and 100°, we may consider the actual gain as only 35 horse-power, or the equivalent of 1,545 pounds of water evaporated. The gain in economy is now with the condensing-engine, until such time as the steam required for the warming of the building reaches 1,545 pounds in weight. The condensed steam or water required for the space you mention will be about 500 pounds per hour, but could all the condensed steam, which would leave the engine when using high pressure to its fullest capacity, be utilized, it would warm in average buildings nearly 2,000,000 cubic feet.

We assume your question to be hypothetical, as your boilers are evidently small for the duty.

HEAT OF STEAM FOR DIFFERENT CONDITIONS.

Q. A DISPUTE has arisen between two local engineers and myself with regard to the value of steam for heating under different conditions, and, as we cannot agree, we have decided to refer the matter to you for a decision. A claims that one-pound weight of low-pressure steam, say at 5 to 10 pounds above atmosphere, will warm more air when condensed to water than if the steam was high-pressure, 50 or 60 pounds. B claims that the high-pressure steam, being the hottest, must be able to warm more air; and I claim that as "*the heat of steam is the same for all pressures*," there can be no difference. Who is right?

A. If a pound of steam at seven pounds pressure above atmosphere is condensed to water at the same temperature as the steam (232° Fah.), 952 units of heat will be realized. If, on the other hand, steam at 60 pounds pressure is condensed to water at 307° (the temperature of the steam), only 899 units of heat are realized. This is on the supposition that the steam is condensed to the temperature of its water only, and then A is right. But from your letter we cannot say that B takes that view of it, and should he consider, or be of the belief, that the steam in both cases would be cooled to the same temperature —say water at atmosphere—he (B) would be right; as in that case the units of heat from 60 pounds pressure of steam to water at 212° will be 985 per pound of steam; while steam in cooling from seven pounds to the same temperature gives off but 972 units.

With regard to yourself, the heat of steam is not the same for all pressures, as at 200 pounds per square inch the *total heat* of steam is very nearly 1,200 heat-units, counting from the freezing point, whereas with steam at atmosphere, or a pound above it, 1,147 heat-units is all that can be realized from it by cooling it to 32° Fah.

SUPERHEATING STEAM BY THE USE OF COILS.

Q. You would do me a great favor if you could answer the following questions: (1) Can I superheat steam to 400° Fah. from a boiler at 70 pounds pressure on steam-gauge by passing it through a coil, 4½ feet long, with cast-iron return-bends, ten pieces of 1-inch pipe in the coil? (2) Would it be safe to risk this coil in a hot fire and let the steam on it at above pressure? Please state the best and safest way to get above degree of heat in the steam, and oblige.

A. (1) In our estimation you can, but the success of your apparatus will depend entirely on the heat of your fire and the quantity of steam you may pass in a given time.

(2) If your coil is exposed to the direct action of a fire, the probability is that it will *burn out*, no matter how much steam you pass through it, and your steam will be heated *above* 400° Fah.

If you find by experiment exactly the length of coil that will be necessary to heat your steam to 400° Fah., you must always have the given or fixed quantity of steam passing the coil. Should you pass less steam, it will be heated above the required standard and the coil endangered, and should you pass more steam the temperature will become reduced. If you wish to make a permanent success of a *superheater*, place the coil in a part of the furnace or flues where the heat is from 50 to 100 degrees greater than the temperature you wish the superheated steam to have, and make the coil of such length by experiment as to give the desired heat when the full quantity is in use. This will prevent the burning of your coil, but it will not prevent the steam from becoming 50 degrees or so warmer when you are drawing it slowly through the coil.

EFFECT OF USING A SMALL EXHAUST AS A HEATING COIL.

Q. I wish to ask of you whether it would be confining an exhaust too much if after running 112 feet from the engine it were turned into coils whose area would only equal seven-tenths of the area of the exhaust at the engine, and would it do to use ¾-inch pipe in the same, or in the coils?

My reason in inquiring about the size of pipe is that the owner has a lot of ¾-inch pipe, and says if he cannot use them here he will have no other use for them, and will consequently have them on hand. I want to use a 1¼-inch pipe.

Again, the steam, I think, will be sufficient to fill them, but I would ask how much 1¼-inch pipe ought the exhaust to fill from an engine whose cylinder is 10½ inches in diameter and 24 inches in length, making eighty revolutions per minute, steam 40-pound pressure.

A. It will throw a back-pressure on your engine, but not sufficient to counteract the gain due to condensing the exhaust steam in a heating-apparatus if you can condense it all or a large portion of it.

Of course, if the engine is now worked up to its full capacity and there is no power to spare, you must not increase your back-pressure. Three-quarter inch pipe will do for exhaust steam if the coils used are header-coils and are short—say not longer than thirty feet. It also appears to us that you can increase the number of pipes in height in these coils, and get the full area of the exhaust. We must not be understood as indorsing ¾-inch pipe for exhaust-steam work, as 1-inch or 1¼-inch is better; but if we had some pipe on hand, with no other use for it, we would design coils from it in which we would use exhaust steam. Whether you will have exhaust steam enough to fill them or not depends on how much pipe-surface you will use.

If you carry steam full stroke in a 10½ x 24 cylinder, pressure 40 pounds, revolutions 80, you will pass about 500 pounds of steam into your engine in an hour. To condense this at exhaust pressure you would require between 1,500 and 2,000 square feet of average heating-surface.

EXPLOSION OF A STEAM-TABLE.

(From the Harrisburg Telegram, December 13, 1882.)

THIS morning, about quarter past seven, a singular accident occurred in the kitchen at the Lochiel Hotel. One of the adjuncts of the kitchen is a long table, the top of which is hollow and contains spaces on which dishes of meat, etc., are placed to keep them warm. The table is heated with steam, which is forced through it from the engine and escapes at one end. This morning the escape was shut off by some means entirely inexplicable, and the steam was forced into the table until it could hold no more, in consequence of which there was an explosion, accompanied by a loud report. The seams of the table were forced open, the legs twisted and bent, and the whole room filled with steam. Dishes were thrown into the air, and things were scattered about promiscuously, but fortunately no one was near enough to be injured. The damage was instantly repaired, and things moved along smoothly in a short time.

[Improvised steam-tables are generally unsafe. All steam-tables have large flat surfaces, which we believe to be seldom braced or stayed, the stiffness of the metal of the top and bottom being depended on, reinforced, perhaps, by a rib. No engineer will construct a water-leg or any other flat surface for a boiler, say 30″ x 60″, and not brace it, and yet they will rig up the most flimsy contrivances for purposes such as above.—ED.]

EXPLOSION OF A STEAM-TABLE.

Q. I CANNOT quite endorse what you say as to the non-staying of steam-pads, such as used at the Lochiel Hotel (*Sanitary Engineer*, Vol. VII., page 148).

I have for years constructed similar fittings, and in scarcely any case is it necessary to stay them, being one-half inch thick, and the steam admitted at three pounds at the greatest, and the condense full open three-quarter inch; little or no pressure is likely to be felt on the surfaces, as the steam being admitted at so low a pressure is condensed almost before it passes through the table. But on "no" account in these fittings should the stop-valve be placed in the condense-pipe, as is, I am sorry to say, often done, with the false notion that more heat is obtained. This, perhaps, was the cause of the explosion at the above hotel.

I see the steam was from the engine condense. Now this is unfair to the engine, as it must tend to give a back-pressure to the engine.

What I do if I get steam for cooking and serving purposes from the engine-boiler, say at 40 to 50 pounds, is to fix a stop-valve at A, Figure 39, with a gauge-glass above it at B, and a safety-valve at C.

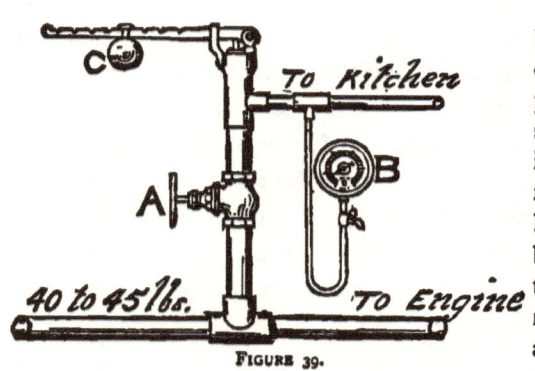

FIGURE 39.

It will be seen that by closing A the steam can be regulated to any pressure on the glass B, say ten pounds, or what is required in the boiler-room to give five in the kitchen apartment. If by any chance it gets over ten the safety-valve relieves it, and so prevents accidents which would otherwise occur.

Reducing-valves, however good, when once fixed, are seldom or never touched, whereas this small contrivance speaks for itself.

A. The arrangement described above will do if one has a weak steam-table which he will not abandon for a strong and properly stayed one, *provided there is no valve on the waste* or condensed-water pipe. But it must also be attended with a comparatively great loss or waste of steam.

It all depends on the attention given to the valve A. If A is not opened sufficiently the cook soon knows it, because his table is not hot enough to suit him; but he cannot tell when he has the maximum heat (without wasting), the return being open.

The notion that the heat is *obtained* by a valve in the waste-pipe *is false*, but that heat can be *retained* by it *is correct*, and, furthermore, the temperatures can be increased by it (by increasing the pressure), which is often desirable.

But the real question is, Why construct any apparatus so frail (when it is possible to do otherwise) that special contrivances, liable to get out of order, have to be arranged to prevent its bursting?

CUTTING NIPPLES AND BENDING PIPES.

CUTTING LARGE NIPPLES.

Q. A COUPLE of years ago steam-fitters came here from New York to do some work, and while here required some 4-inch close nipples, which they succeeded in cutting without the aid of a lathe or pipe-cutting machine; their only tools being large stocks and dies that would cut up to 4-inch, and "diamond points"—*i. e.*, steel chisels with peculiar-shaped points—and, of course, a vise.

How to cut small close nipples we know, when there is an opportunity of passing the "stocks" over the coupling which holds the nipple, but with large sizes, especially 4-inch, as that is the limit of the guide-bushing, we cannot see how the cutting can conveniently and practically be done, and if done at all, how it can be *straight*, as it is not in the power of a man or men to catch a 4-inch thread straight by main force with the die turned.

An explanation will oblige *two* young steam-fitters.

A. "Thermus" sends us the following explanation and diagram. (Figure 40):

It is presumed that the steam-fitter has a 4-inch stock and die A, a vise B, a pair of 4-inch tongs D, and a promiscuous assortment of pipe and couplings.

In the vise B fix a convenient piece of 4-inch pipe G, on which there must be the coupling F, which forms the nipple-chuck; on the other end may be the coupling I. Into the coupling F screw the (half-cut) close nipple. Within the 4-inch pipe G slip a piece of 3½-inch pipe H, until the coupling J (3½-inch) comes against the coupling I. Then reverse the die-plate *a* from its usual position in the stock and pass it over the 3½-inch pipe, bringing the large side of the die against the nipple to be threaded. The lead-screw and guide C is to be then run inward and centered on the pipe H by the set-screws, or the 3½-inch guide-bushings may be used if there are not enough set-screws to properly adjust the centre. When that is done, prevent the coupling J from revolving by a pair of tongs, and revolve the die-stocks in the usual manner.

This will run the stocks off the guide and lead-screw and force the die on the pipe *straight*.

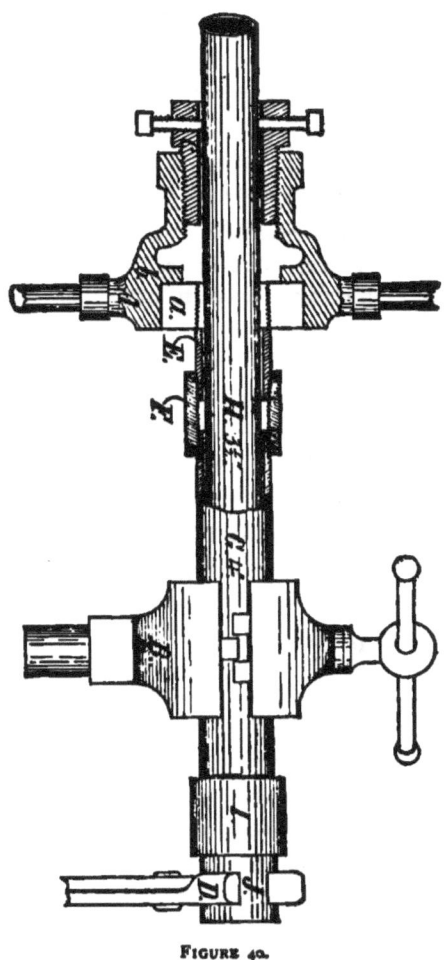

FIGURE 40.

When a couple of full turns are taken and the die has "caught," slack up all and withdraw the 3½-inch pipe so as to cut the remainder of the thread without unnecessary friction.

If one is going to make a business of cutting 4-inch short nipples in this way he must provide himself with a pipe G, in which the thread will be long enough to run through the coupling and meet the end of the nipple E, to prevent the latter from running into the coupling, but for once or twice a common piece will do.

CUTTING CROOKED THREADS.

Q. How can I cut a *crooked* thread on a close nipple? I frequently require such pieces, but have to bend them hot in a fire, which spoils the two couplings I have to hold them in.

A. "Thermus" replies that his method is shown in the accompanying illustration, Figure 41.

Secure a piece of pipe G, say four inches, in the vise, with the coupling F for a "nipple-chuck." Insert the half-cut nipple E. Then through the centre of all affix the pipe H, say 2½-inch pipe, using wedges $w\ w\ w\ w$, to hold it approximately true. At the end use a flange K, or anything which will keep the pipe H from pulling through. Then apply the die a turned in the stocks—*i. e.*, with the largest side outward with relation to the stocks. Then, if the lead-screw b, into which the guide-bushing c (2½ or 3 inches, according to the amount of eccentricity required) is fitted, is run inward and fastened as shown, by revolving the die and holding the centre guide H from revolving a crooked thread may be started. When the die is fast on the thread, so as not to strip, the whole may be slackened and removed, and the thread finished without unnecessary friction on the lead-screw.

Different degrees of eccentricity can be obtained if the lead-screw is fitted with *three* or more set-screws, otherwise the bushings or wooden wedges will have to be depended on.

The drawing will suggest other modifications to the practical man.

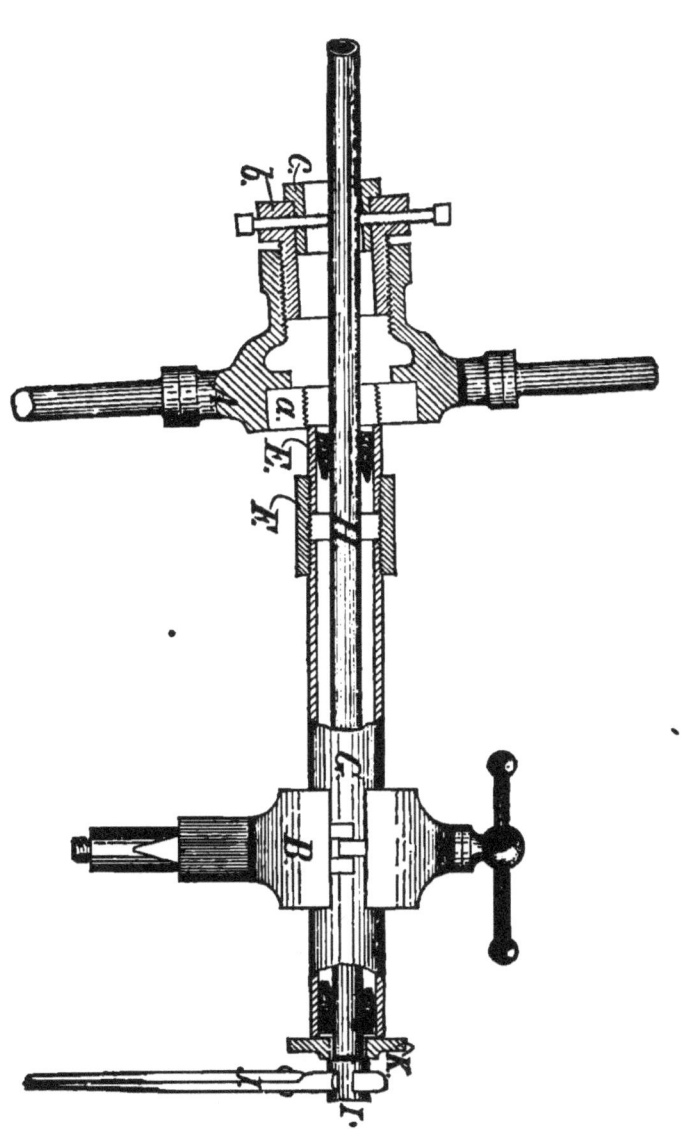

FIGURE 41.

GETTING A CLOSE NIPPLE OUT OF A COUPLING AFTER A THREAD IS CUT.

Q. I NOTICE "Thermus's" explanation of how a 4-inch close nipple may be cut with ordinary stocks and dies, but I would like to ask him what use a 4-inch close nipple, or any close nipple, is to a steam-fitter when it is forced as tightly within a common coupling as it will naturally be when it has to give resistance enough by friction on the threads to force the uncut end of the nipple into the die? My experience has been, that there is not one chance in ten of removing it without spoiling it, as the pressure on the tongs will cut into it and make it oval, as well as mar the threads within the hook of the tongs.

A. "Thermus" sends the following reply:

"Nipple-cutting is looked on in the pipe-shop as lead-trap making was a few years ago by the plumbers—as very good work to keep the boys at when there is nothing doing outside—and is so much detested by a good workman that he would generally go home if there was nothing else for him to do.

"But with smooth, beautifully made traps, etc., have come machine-made nipples of all lengths and sizes, which can be bought for very little more than the same length of pipe, were the nipples put end to end. This, of course, put an end in a great measure to cutting nipples in jobbing shops, and will probably account for our correspondent not knowing how to get a close nipple out of a common coupling. But it was not, I am sure, to make close nipples for the trade with a stock and die that former correspondents wanted the information; but to be able to cut one or two such nipples when they wanted them badly and could not wait to send to a large city for them.

"Such a method may be called a 'trick of the trade,' or a finishing touch to the piper's education, and it is legitimate, though perhaps not regular. How to get a close nipple out of a coupling without getting it out of shape is another 'trick,' or it may be *several* of them.

"After removing the die from the nipple, it is presumed that some kind of fitting will be used in connection with the nipple, and that the screwing of the fitting on to the nipple before it is removed from the coupling will not prevent its being used where it is required. If so, clean the thread, lead it, and screw the fitting on (say it is an elbow) until it is sufficiently tight on the thread, or until the nipple begins to screw into the coupling further. Then hammer the coupling—not heavily—keeping a strain on the elbow in a direction as if you were going to unscrew it. After a few light blows or so are struck the nipple will unscrew from the coupling easily, but will remain fast in the fitting in which it is going to be used.

"If the nipple must be taken from the coupling without having a fitting on it, run a lock-nut over the thread on the nipple, then screw a coupling on the nipple no tighter than it can be removed without spoiling the latter; the lock-nut can be then brought against the end of the coupling to form a 'jam-nut,' when the other coupling may be hammered, as before explained, and the nipple removed by using a tongs on the second coupling. By loosening the jam-nut, the second coupling may then be removed."

BENDING PIPE.

Q. We have occasion frequently to bend iron gas and steam pipe. As a general thing the bends are unsightly, either flattening the pipe or drawing it thin on the back. I have tried filling the pipe with sand, but I cannot see that it improves matters. Is there any simple practical method in use which will give a uniform bend? An answer through your paper will greatly oblige a constant reader.

A. We have seen bending-machines for this purpose, but we cannot say whether or not they can be purchased in the market. We think they have been mostly home-made. They consist principally of a lever with a grooved wheel, with other grooved wheels of different diameters, around which the pipe is bent.

The ordinary method of the steam-fitter is to bend his pipe in a vise (without filling), in the manner shown in Figure 42. In a little time, with practice, he usually succeeds in making a bend on most sizes of pipe below three inches in diameter that will not be so much out of shape as to attract attention.

The method is to heat the pipe to be bent the whole length of the bend, if possible, so as to complete the operation at once. A parallel-jawed vise must be used which is sufficiently sharp in the serrations of the jaws to prevent the hot pipe from being drawn from it when pressure is applied in the direction of the arrows. At the same time the pressure exerted by the vise-jaws must not be sufficient to flatten the pipe in the direction of the grasp of the jaws—simply to hold it.

If, now, force is applied to the pipe in the direction of the arrows, the pipe begins to bend, and the tendency of the warm part of the pipe (within the vise-jaws) is to spread in the direction of the grasp of the jaws; but being prevented by the jaws from becoming any wider than the diameter of the pipe, it, as we may say, prevents the arch from spreading. This makes the parts which touch the vise a neutral line, forcing the inside of the pipe to compress and the outside to elongate, at the same time keeping the pipe practically round at all parts of the bend.

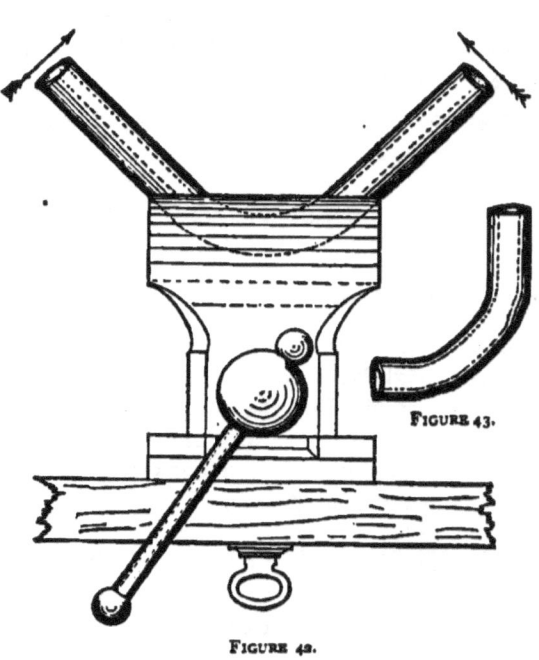

FIGURE 43.

FIGURE 42.

If a pipe has been warmed too much of its length and cooled with water it is likely to pull *thin* on the back, as shown in Figure 43, and be irregular in the radius of the bend.

Different qualities of iron act differently in bending, and a failure at first should not discourage the beginner. The neutral line of the bend which touches the vise becomes chilled and compels the stretching of the back of the bend and the upsetting of the inner side of the pipe. For this reason it is usually better to bend the pipe toward its coldest side (having the cold side uppermost in the vise), to prevent wrinkling the inside of the bend.

All pipe up to and including 1½-inch may be bent in this way. Two-inch is more difficult when the radius is short, on account of the thinness of the pipe compared with its diameter. Two and one-half inch pipe bends better than two-inch, the radius being proportional to the diameter of the pipe. All pipes below 1½-inch should bend properly to a quarter turn or less, with a radius equal to twice their diameters.

CUTTING LARGE NIPPLES.

Q. My way of cutting 4-inch close nipples, or any other size, is to have a long thread on a piece of pipe. I back the coupling of the long thread sufficiently to let my short nipple go far enough not to rupture the thread, then I run the coupling on my long thread—that is, when I remove my die; then I unscrew the coupling from the long thread until the ends are separated, when I generally remove the nipple with my hand.

A. Your method is the ordinary one of removing a short or close nipple from a nipple-chuck. It is assumed that it is an easier task to remove one or two 4-inch nipples from an ordinary coupling and short piece than to make a 4-inch thread six inches long on the piece, for the purpose of making a nipple-holder or chuck.

CUTTING VARIOUS SIZES OF THREADS WITH A SOLID DIE.

Q. I work in a gas-fitting shop and help a fitter. The dies we use are the ordinary solid ones and cannot be made smaller or larger. Now and then it happens that the fittings are a little too small for the threads which we cut on the pipe, but the fitter has a way of making the thread on the pipe small also, so as to fit the fittings. He always sends the boys out while he is doing it, as he says it is a "trick of the trade," and should not be shown to any boy not an apprentice. He is an old-countryman, and had to serve an apprenticeship himself, he says. I told him I would find out, and take the liberty to ask you.

A. He probably uses a piece of sheet-metal—tin, sheet-iron, brass, or copper will do—placing it over the cutting-edges of the die at one side only. However, if this is not his way, such a method can be used for the same purpose, and we think it is no secret in old-fashioned machine-shops, at least when applied to a tap, for a hole may be made larger than its tap by the same process.

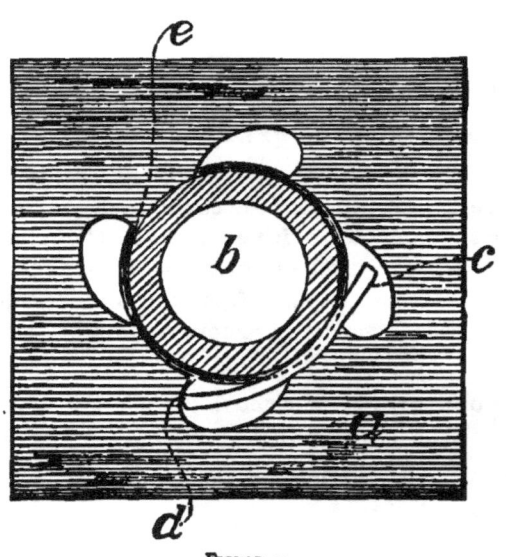

FIGURE 44.

Let *a*, Figure 44, be a common solid die and *b* a pipe which has been already threaded by it. Then take a piece of thin soft metal *c* and place it over one cutting-edge of the die, one end abutting at *d*. Force the die on again in the usual manner. The result is that the cutting-edge of the die *e* is drawn into the pipe the thickness of the slip of metal—"chasing" as it were a thread already cut to a smaller size in diameter. The thickness of the slip determines the amonnt of reduction.

In the same way when a hole is made with a tap, and if it is necessary to make it a little larger, run a strip of copper or tin down one side of the tap. The points of the cutting-edge will stick in it and carry it around the hole, forcing the cutters of the opposite side into the work.

In the answer preceding, in our number of November 8, 1883, we omitted the cut, and though the reply is clear enough to a technical reader, fearing the younger members of the workshop may not be sufficiently benefited by it, we give it in this issue. The thin piece of metal c, or it may be two or three thicknesses of tin plate, is set in the die after it has already cut a thread on the pipe. The die is then again forced on the pipe, and as the widest side of the die is toward the point or narrowest part of the pipe-thread, it readily catches. The result is then the deepening of the threads by the cutting-edge e, which is drawn in the direction of the centre of the pipe. The operation may be repeated for a further reduction of the thread by adding more strips. A change of taper of the thread (not accurate) may be obtained by the same method by running the die on only part way.

RAISING WATER AUTOMATICALLY.

CONTRIVANCE FOR RAISING WATER IN HIGH BUILDINGS.

MR. G. STUMPF, a civil engineer in Berlin, has recently devised and advocated a contrivance for raising water into the upper parts of high buildings, to be used principally for fire purposes, but serving also other uses in the upper stories. Figure 45 illustrates the apparatus. The water is admitted from the street-main into the pipe K. By opening the stop H the water rises through the pipe F into the tank A (which is air-tight) until it

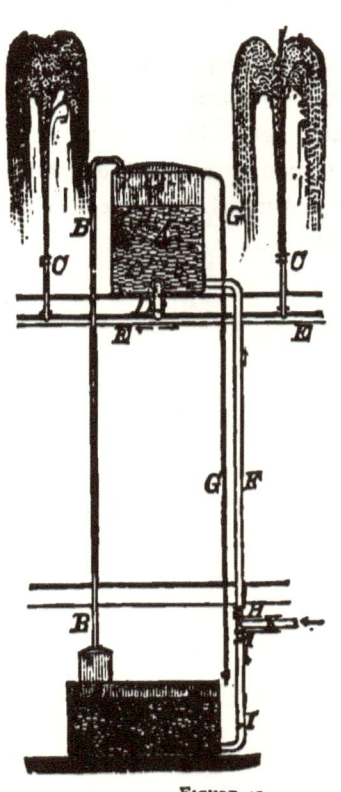

FIGURE 45.

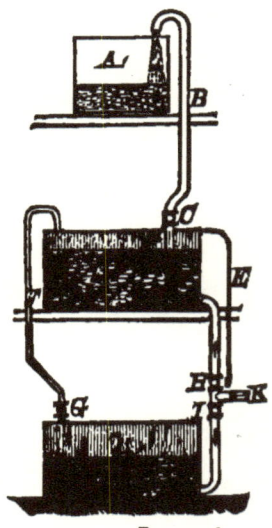

FIGURE 46.

is filled, which fact is indicated by the overflow-pipe G; thereupon the stop H is closed and the stop I opened. The water from the main enters the tank L, which is also air-tight, and in doing so compresses the air and forces it through the pipe B into the tank A. The water in the latter is, therefore, under the same pressure as the water in the tank L. From A the water is then permitted, by opening the stop D, to enter the pipe E and to flow out at C under a very much greater pressure, and consequently rises to a much greater elevation than if there was a direct connection with the street-main. Of course the flow of water from C continues only until the tank A is emptied, and its refilling must be attended to as described at first.

The quantity of water at once available for fire or other purposes is, therefore, dependent on the size of the tanks A and L.

There are instances, particularly in the country, where a good deal of water may be had, but with very little pressure. In this case an apparatus like Figure 46 may be used. The air-tight tank D is filled from the supply-pipe K until the water runs out through the waste E. The stops in the latter and the stops H and C are then closed and the stop I opened. The water from the supply-pipe then runs into the air-tight tank L, compresses the air which is forced through the pipe F into the tank D, and raises the water contained therein through the pipe B into the tank A situated

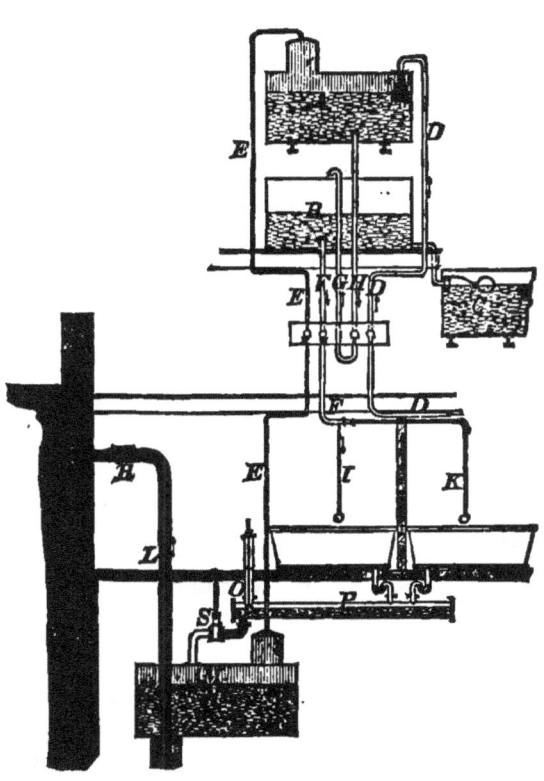

FIGURE 47.

at the top of the house, from which the supply is then taken to the various parts of the building. The tank D upon being emptied in this manner must be filled as before.

A further application of this principle is shown in Figure 47. The water from the main enters through the pipe D and fills the air-tight tank A, compressing the air which is forced through the pipe E into the air-tight tank M in the cellar, and forcing whatever water or sewage which may have collected therein from the cellar into the pipe L, which delivers into the sewer, which in this case is supposed to be higher than the cellar-floor. Back-flows from the sewer is prevented by a flap-trap, R. The tank B, which is not air-tight, is filled from the tank A in a manner readily noticed, and from this the various cisterns and hydrants in the house are supplied.

There is no question that Mr. Stumpf's contrivance would be useful under certain conditions. A disadvantage lies in the fact that it is necessary to see to the filling of the upper tank whenever it becomes empty, as no automatic arrangement for doing it is given.

APPARATUS FOR RAISING WATER.

A CORRESPONDENT writes : "I see cuts and explanation of a 'contrivance for raising water in high buildings,' said to be devised by Mr. G. Stumpf, of Berlin. The device as shown in the cuts is not complete. I see nothing by which the tank L can be emptied, and this tank must necessarily be emptied whenever it gets full, as you or any intelligent reader can see. Inclosed you will find the sketch of an apparatus which I invented three or four years ago, which is automatic in its action. (See Figure 48.) It will work with the smallest possible quantity of water.

"The sketch explains itself. A constant stream of water runs into the chamber A. Suppose both tanks are empty, and the water turned on. The water runs through the supply in bottom of chamber A to top tank through the check-valve until the water rises to level of overflow. Immediately on its doing so the check-valve closes. The overflow supplies bottom tank, which, when filling, compresses the air above the water. The compressed air passes up through the air-pipe into the top tank, pressing on the surface of the water in the tank, and forcing the water up through the outlet. When the lower tank is full the float and

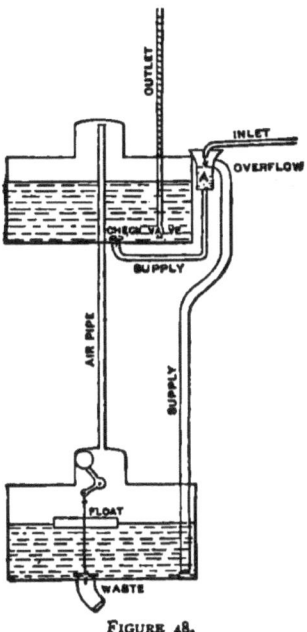

FIGURE 48.

weight open a valve in the bottom, and allows it to empty. While it is emptying the top tank is filling up again. With this apparatus water can be raised any height, all that is necessary being to multiply the tanks above one another and connect the air-pipe to each."

MOISTURE ON WALLS, ETC.

THE CAUSE AND PREVENTION OF MOISTURE ON WALLS.

Q. I HAVE a problem that I would like solved. We have a country house on the Mississippi bluffs. The house is built of rough stone, the foundation-walls being three feet thick and the height of the basement ten feet. Above this the walls have a thickness of two and a half feet. The plaster was put on the stone without any lathing, and the consequence is that the house is so damp that water sometimes runs down the wall-paper in small streams. By what system can we get rid of that dampness, which is, I am sure, as pernicious to health as to comfort?

It would be a great deal of trouble to have the whole house plastered, and I suppose there must be some easier way to remedy such a defect. If you will advise us you will greatly oblige.

A. The dampness complained of is caused by the condensation of vapor from the air upon the surface of the walls, just as it is condensed upon the outside of a tumbler when filled with ice-water in warm weather. It occurs when the air is well charged with moisture, as it generally is during summer weather, and at such times as the house-walls happen to be cooler than the air. Such conditions often exist in our climate when a warm day follows a cool one. Air is capable of sustaining watery vapor in an invisible form in quantities varying directly with its temperature. The quantity so taken up and held is nearly doubled with every increase of 20° Fah., provided water is exposed to such air for evaporation. Thus, air at 60° Fah., if water or moist surface had been exposed to it, contains 5.77 grains of water per cubic foot, while air at 80° under like conditions contains 10.98 grains per cubic foot, or 5.21 grains additional. If air in the last condition be chilled 20° by contact with any cool substance, this 5.21

grains of water per cubic foot of air is at once deposited on such cool surface in the form of dew. If the air be heated by fires within the house and no water exposed to it, then its subsequent cooling, when brought in contact with the cooler walls, produces no condensation. But if after the walls have been cooled by a northerly wind for a few days we have a warm wind from the south fully supplied with moisture, as is often the case in summer, the air is chilled at once below the dew point when it comes in contact with the cooler walls, and condensation or deposit of water ensues upon their surfaces.

There are two ways to remedy the trouble: First, by building a fire in the house, by which the walls may be artificially warmed, which, though efficient, may not be conducive to comfort in summer; second, by covering the inner surfaces of the walls of the house with a lining which is a non-conductor of heat. This will prevent the rapid transfer of heat from the air to the walls, which will become more slowly heated from the outside. Many stone buildings in the Old World are lined with tapestry hangings, which are tolerably successful in checking condensation, but these accumulate dust and insects to a degree that renders them disagreeable. Modern practice has discarded them in favor of a confined air-chamber between the wall and the plastering; though we understand the mistaken practice that you describe is still quite common in the Western States. For this purpose the plastering is spread upon laths instead of directly on the walls; and in order to secure a space of one or two inches between the back of the lathing and the walls, a furring or strip of board is attached to the wall at such intervals as to give proper nailings for the laths. For stone walls these furrings should be not less than two inches in thickness.

Such air-spaces are now almost universal in our brick and stone buildings in the East. They are sometimes, however, helps to the rapid spread of fire, when no care is taken to interrupt them at the several floors. The proper way is to fill this space with mortar or brick-work at every floor, several inches in thickness, so as to effectually cut off all communication from one story to another through such air-spaces. If the partitions are made of brick or stone they should be furred and plastered in the same way. Wooden sheathing or panel-work is as good as plaster, if furred off two inches from the wall, but plaster is generally used on account of its cheapness.

Walls are sometimes built with an air-space within them, but if lined with brick or stone inside this space, the condensation will still occur when the air is being cooled until that part of the wall inside the air-space has been warmed up to the same temperature as that of the air.

The thicker the walls of a house happen to be the longer time does it take to warm them up, so that thick walls, such as you describe, are more troublesome in condensing moisture than thin ones of brick, the latter being more readily warmed.

EFFECT OF MOISTURE ON SENSIBLE TEMPERATURE.

Q. WHAT will be the result of running a heating apparatus, as regards the effect of the heat, with and without evaporation? In other words, would a room heated to 60° Fah. feel the warmer to its occupants if the air were dry or if it were moistened with watery vapor from evaporations? I do not know whether I have made my meaning plain, but my question arises from having noticed in summer time that on a moist day, or what is called "muggy" weather, the heat is felt more than it is on a dry day with a higher temperature as shown on the thermometer.

A. The presence of moisture in the air has a strong influence on bodily sensations as regards temperature. In an atmosphere nearly saturated with moisture, as in the west and south of England, Ireland, or Normandy, 60° Fah. is sensibly as warm as 75° Fah. would be in Canada or Minnesota, where the air is comparatively dry.

Were it possible, therefore, to maintain in a room artificially warmed to 60° Fah. a nearly saturated condition of the air as regards moisture, such a room would be as comfortable as one heated to 70° Fah. which was nearly free from moisture. But with the external air at the freezing point in this country it is practically impossible to supply the vapor required to maintain such moisture; and it would take more fuel to vaporize the water than it would to heat the room. The only way to effect it would be to have a room absolutely air-tight and without ventilation; 5.46 grains of water to each cubic foot of air would be required; 4.02 grains of this must be evaporated by heat, requiring 0.612 units of heat, while the amount of heat necessary to heat a cubic foot of air from $32°$ to $68°$ is 0.635 units, or very little more than that required to effect the evaporation.

MISCELLANEOUS.

HEATING WATER IN LARGE TANKS.

Q. A CIRCULAR tank with flaring sides, measuring 4' 3" on bottom, 3' 7" on top, and 3' 8" perpendicular height, is used here by laundry for washing purposes. It stands on roof and is housed in. It is supplied by city water, which runs in over top, connected with ball and cock. At present it is heated by steam. I wish to do away with steam and heat by stove and coil, or some other hot-water apparatus. From the bottom of the tank to the floor directly beneath upon which the stove will stand the distance is 13' 6". What or how much heating-surface in the coil is needed to heat this tank full twice a day to 200° F.?

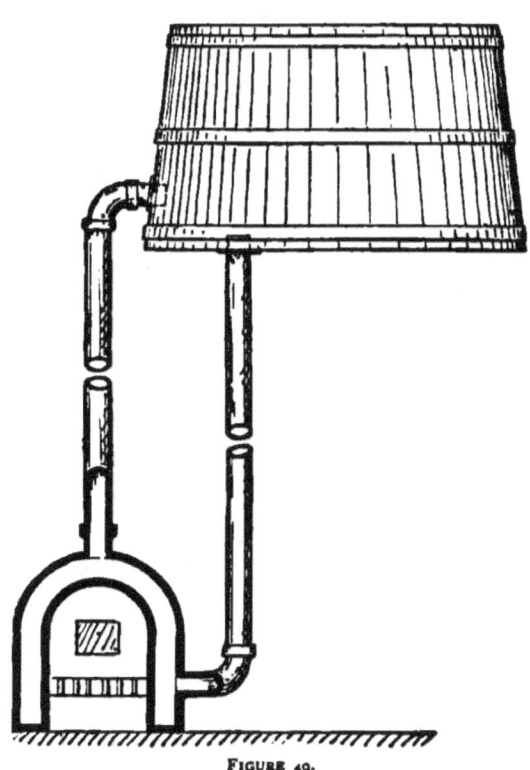

FIGURE 49.

A. Your tank will hold about 2,800 pounds of water, which, if warmed from 40° to 200° F. twice in ten hours, will require 896,000 *heat units*, or 89,600 *heat units* per hour—equivalent to the evaporation of about 1½ cubic feet of water per hour. A green-house boiler of from 25 to 30 square feet of surface will do. Connect it as shown in the sketch, Figure 49.

HEATING WATER FOR LARGE INSTITUTIONS.

Q. In a public institution using a large quantity of hot water it is proposed to have a hot tank in the garret, and to use for heating the water in this tank the steam-boiler which heats the building with steam in the winter, a coil being placed in the tank. In summer it is proposed to heat the tank by a small auxiliary furnace in the garret. Will you please give me some directions about the arrangements common in such cases? Also whether, as fixtures will be connected with a pipe descending from this tank, it will not be impossible to obtain hot water until the cold water which gathers in the pipe by its cooling has been drawn out, for I suppose you cannot get a circulation in a case like this.

On this account I have recommended to the parties that they had better heat their water in a closed tank in the basement, letting it rise to the reservoir in the top, and then return to the closed tank, thus always having a circulation.

A. The sketch (Figure 50) shows how we would

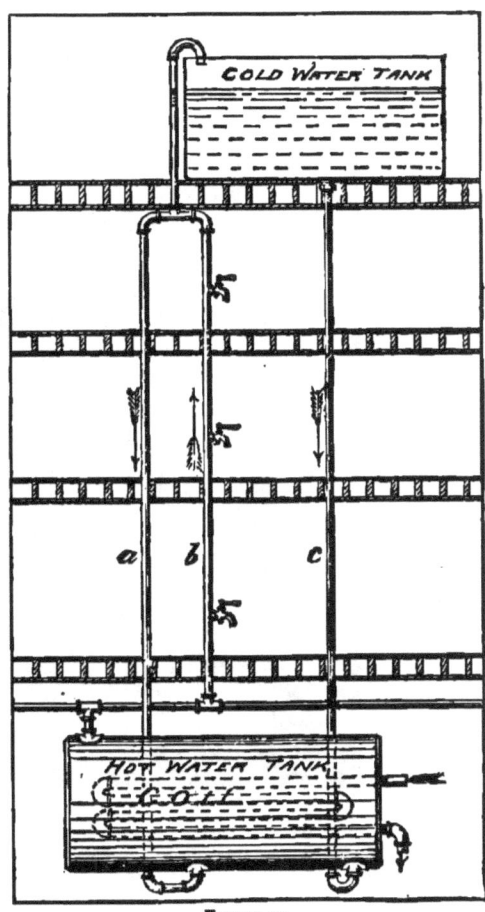

FIGURE 50.

do this in New York, especially in a high building. The water may be pumped to the cold tank, or should the water-works pressure be great enough, it may be regulated by a ball-cock. From thence it will run to the hot tank, and will level up again to the same height in the distributing and circulating pipes *a* and *b*. The small pipe over the top of cold tank is an air-vent. With this arrangement the water will always be warm at the faucet.

The coil within the lower and closed tank may be 1-inch pipe for live steam or large pipe for exhaust steam. Brass pipe is generally

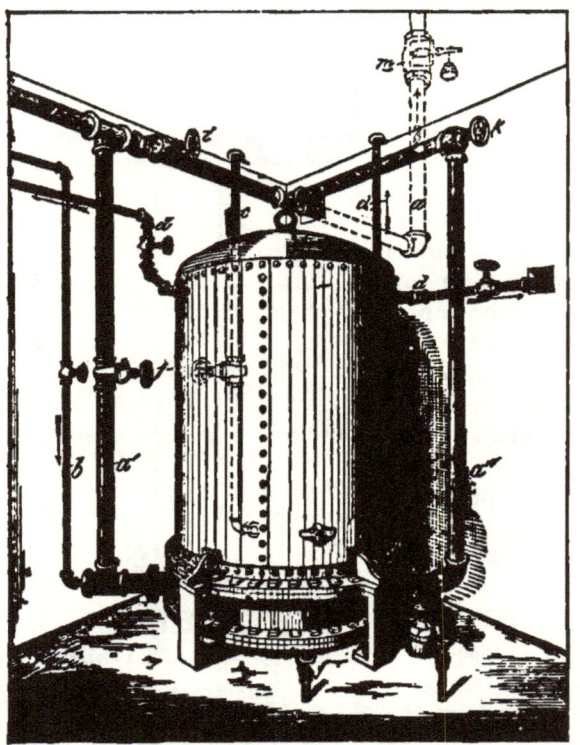

FIGURE 51.

used. With the hot tank in the top of the house the water will not circulate and be warm. The result will be as you state.

You also say it is proposed to take steam from the heating-pipes to warm this tank. There are heating systems you might tap in this way and not spoil the circulation, but on general principles it is wrong.

Figure 51 shows the method of making hot water for distribution within the Hotel Warren in Boston. A Berryman heater of special construction is used, and the exhaust steam from the elevator pumping-service is utilized. The pipe *a* leads directly from the pumping-engine. When the valve *i* is closed and the valves *j* and *k* open, the passage of the exhaust steam is through the pipe *a* to the pipe *a'*, thence through the tubes of the heater, the uncondensed part passing off by the pipe *a"*, thence entering the pipe *a* again beyond the valve *i*, and either passing to the roof in summer time or into the heating-coils in winter, by the resistance of the back-pressure valve *m*, which is then loaded to exceed the pressure to be carried in the house-heating apparatus. The pipe *b* is for the admission of high-pressure steam from the boilers, should the exhaust be not in use. The valves *j* and *k* are then closed and the water of condensation is taken care of by the trap *c*. The pipe *f* is a sediment-pipe for the flushing out of the water-chamber at the base of the circulating-tubes within the heater. The pipe C supplies the heater with cold water, and the pipes *d* and *d* are the distributing. The circulating-pipes return to the heater parallel to the pipes *d* and enter near the bottom.

QUESTIONS RELATING TO WATER-TANKS.

Q. How MUCH water will a tank hold if it is 45 feet long by 27 feet wide, 6 feet 6 inches deep at one end, with a regular incline to a depth of three feet at the other end? How large a pipe will it require to let in water enough to fill it in five hours, under a constant head or pressure of 40 pounds per inch? How much water at a temperature of 190° Fah. will be required to raise the contents of the tank to 70° Fah. from 36° Fah.? What is the best and most economical means of heating and keeping water in the tank at a temperature of 70° Fah., steam not being available?

A. The tank will contain 5,163 cubic feet, or 38,619 U. S. gallons. This is equivalent to admitting 129 U. S. gallons per minute, and this quantity will pass through a short 2½-inch pipe at the pressure you state.

If your pipe has short turns, and is of any considerable length of ordinary pipe and fittings, make it three inches in diameter, with straight-way valves.

Thirty thousand gallons of water warmed from 36° to 70° will cool 8,500 gallons from 190° to 70°, the resulting mixture having a temperature of 70° Fah.

To keep the water hot under such conditions, use an ordinary house-heating hot-water boiler, and connect it so the water can circulate to the tank—the greater the difference in level between the boiler and the tank producing the quicker circulation. A boiler of 500 square feet of heating-surface, with 6-inch connections, will heat the tank in about one hour

FAULTY ELEVATOR-PUMP CONNECTIONS.

WE would direct the attention of readers to a method of pump-connections sometimes resorted to by steam-fitters in their endeavors to remove water from the cylinders of elevator-pumps, a sketch of which has been furnished by a correspondent. (See Figure 52.)

It is well known to the engineer that elevator-pumps are regulated to start and stop automatically, as the level of the water in the tanks requires. When the times the pump is not running the water of condensation accumulates in the steam-pipe back of the "chronometer-valve." When the pump starts again it is desirable to remove this water as quickly as possible and to do it automatically, to which end a "pot" steam-trap is generally used. Some engineers attach their traps to the steam-pipe close to the steam-chest, as shown by the dotted lines, so as to receive the water before it goes into the cylinders, but a few, with the hope of removing any water that is condensed in the cylinders, connect the trap in the manner shown by the heavy lines in the sketch.

To have the trap at all operative with this method, the cylinder-cocks c c c have to be left open, but as the trap does not open directly to atmosphere except for a few seconds, at intervals of as many minutes, it has the same pressure in it as there is in the steam-pipe or in the ends of the cylinders which at the moment are taking steam. The result of this is to pass steam from the ends of the cylinders which are taking steam to the ends which are not taking steam, so either wasting it to atmosphere through the exhaust-pipe or causing a back-pressure on the pistons. But this action goes further. The water which passes out through a cock *c* at the end of the cylinder goes largely in the direction of least resistance, and flies into the other end of the same cylinder through the pipe *i*, or back to the other cylinder through the pipe *j*, making abortive the very object for which the trap was originally applied, wasting steam and causing the pump to pound.

If traps must be placed on cylinders, one should be used at each end of each cylinder; but we think that a trap on the steam-pipe, as shown by the dotted lines, with the pipe *d* connected with the pipe *f* outside the trap, or to some other pipe in which there is no pressure, is all that is required. This allows the trap to drain the steam-pipe of water automatically, and gives the engineer the opportunity of draining the cylinders by the cocks *c*, which he can regulate by hand to a fine adjustment.

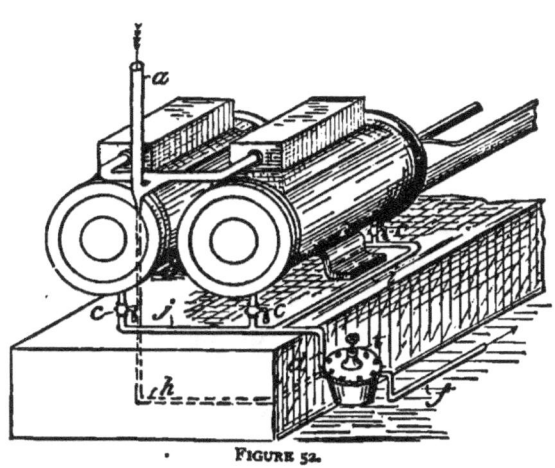

FIGURE 52.

It may be well to state for the benefit of some that the cocks *c* cannot be finely adjusted with the trap attached, as it is necessary to have a full pressure of steam in the traps to discharge it. Otherwise it would be inoperative and useless.

A correspondent writes us in reference to the above subject: "Would it not be well to let the steam-trap remain as it happened to be in the case of the faulty connection, and insert swinging cluck-valves in the pipes *c c c c*? I have been doing that for some time now, and get better results than when I attached the trap to the steam-pipe or steam-chest."

BOOKS ON HEATING SEVERAL BUILDINGS FROM ONE SOURCE.

Q. Is THERE any book on the heating of different buildings from one central station? Or has anything been published in the *Sanitary Engineer* or elsewhere on this subject?

A. This subject is not fully treated of in any work that we are aware of. In chapter XIX. of "Baldwin's Steam-Heating for Buildings" allusion is made to this subject in connection with "scattered buildings heated from one source." If it is proposed to deal with any special case, our advice would be to employ an expert either to improve a defective system or to design a new one, the cost of his services being nearly always more than compensated for by the reduced first cost, not to consider the cost of maintenance as between a defective apparatus and a properly constructed one.

COAL-TAR COATING FOR WATER-PIPES.

Q. WE should like your opinion as to the benefits of coal-tarring water-service pipes inside.

A. Some protection against corrosion is necessary for any iron pipe used for water-supply. Plain iron, whether cast or wrought, is liable to become incrusted, and to both rust out and become obstructed. Dipping the pipe when heated in a hot bath of coal-tar (known as Dr. Angus Smith's process) is effective in preventing such corrosion to a very great extent.

From statistics collected by a committee of the New England Water-Works Association, and published in its proceedings for 1884, it appears that pipe of this kind has been used for a number of years for service-pipes from main to house, with very satisfactory results in Lowell, North Adams, Quincy, and Springfield, Mass., Pawtucket, R. I., and Wilmington, N. C.

In Haverhill, Mass., it has not given satisfaction; in Northampton, Mass., it is stated to give a bad taste to the water for some time; and in Newton, Mass., it is said to last eight to twelve years.

For cast-iron main-pipes, the coal-tar coating which was first introduced into this country by Mr. Kirkwood on the Brooklyn Water-Works in 1857 is used universally with very satisfactory results.

FILTERS FOR FEEDERS OF HOUSE-BOILERS—OTHER MEANS OF CLARIFYING THE WATER.

Q. I send you herewith a rude sketch (Figure 53) of proposed application of the Crocker or the Grant filter, globular, to clean the feed-water supplied through an automatic feeder to the boiler I use in heating my house. The boiler is six feet by two and a half, and of steel. Please submit to best authority. Can you suggest any improvement on the mode employed, or do you see any difficulty? The good filtration of water is a great desideratum. Would a filter lessen the rapidity and quantity of the pipe-delivery of water much? The filters I named are packed with animal charcoal and a punctured or wire strainer.

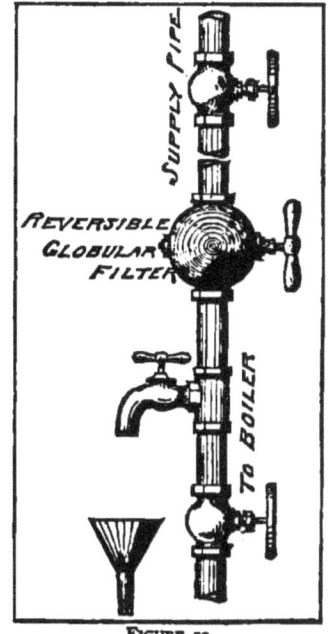

FIGURE 53.

A. 1. Filters of small capacity are of very little use except to arrest fish or large particles, and must be cleaned often.

2. When the gravel, charcoal, wire, or whatever else a filter may be filled with, becomes clogged with the particles arrested, it will retard the flow of water.

Settling the water in an iron holder or reservoir (a kitchen-boiler will do) having the inlet and outlet both in the top is a good method where a comparatively small quantity of water is to be used, as with an automatic feeder for house-boiler.

TESTING GAS-PIPES FOR LEAKS AND MAKING TIGHT JOINTS.

Q. I would like to inquire the best way to test gas-pipes in a building for leaks. Also, how to make perfectly tight joints in said gas-pipes.

A. If the house is in progress of construction, see that all the outlets are carefully closed with caps, and that the foot of the rising line is stopped. Then at any convenient side-light attach the ordinary gas-fitters' pump, which is simply an air-pump. To the same side-light, or an adjacent one, attach the mercury-column gauge used by gas-fitters with a column from fifteen to twenty inches in length. Great care must be now taken to prove that there are no leaks in the gauge or its connections or cock, and in the pump and hose connection, and a good cock should be used between the permanent gas-pipe and any temporary connections to pump, so that it may be closed immediately the pumping stops, to prevent back-leakage of air through the pump-valves or hose-joints.

When all is complete, pump the pipe system in the house full of air until the mercury rises at least twelve inches. Then close the intermediate cock before mentioned, and should the mercury column be found to "stand" for five minutes, it is reasonable to assume that the pipes are sufficiently air and gas tight for any pressure they can afterward be subjected to. But as it is the rule in the most carefully done gas-pipe work to find the mercury will not "stand," as there will be leaks that would escape the most careful workman, it is necessary then to locate them.

Should there prove to be a very large leak, it will be apparent at once, as it will be impossible to get a pressure worth considering, the mercury simply bobbing up and down in the tube.

It may be an outlet that has been neglected to be closed, or it may be a long split in the pipe. If the former, and very close to the pump, the mercury will not respond; but should it be far away, with considerable length of pipe to cause resistance, the mercury will jump and return as suddenly. But should there be a split pipe or an aggregation of small leaks, the mercury will run back steadily, though slower than it rises, between the strokes of the pump. Should it rise well in the glass and sink at the rate of about one inch in five seconds, small leaks only in fittings or joints may then be anticipated. Of course, there are exceptions to these rules, which are only for general guidance.

To locate a leak, then, that cannot be heard blowing, strong soap-water applied with a brush or sponge may be used. The liquid is rubbed over suspected joints or fittings and air-bubbles are blown by the escaping air.

Sometimes it becomes necessary to use ether in the pipes in locating leaks, if the pipes are under floors or in partitions. The ether is put into a bend of the hose or into a cup attached to the pipe and blown into the pipes with the air. By following the lines of the pipes the approximate position of a leak may then be determined by the odor of escaping ether.

In very large work it is well to prove a floor at a time, and when all are done, connect them with the riser and prove as a whole.

The best thing for making pipes tight for coal-gas is gas-fitters' cement, which is a common grade of sealing-wax. The threads of the pipes should be immersed in it when warm and let drain, and the fittings also are sometimes so treated. To put the pipes and fittings together both are warmed and screwed tightly and allowed to cool. Porous places incidental to malleable iron or shrinkage-cracks in malleable-iron fittings are generally stopped with this cement, but a split or crack should never be so mended, as it will be an element of danger.

For naphtha-gases some of the heavy body asphaltum varnishes are considered best, such as black air-drying japan, or black baking japan, but paraffine varnish should not be used. To use the japans both threads of pipes and fittings should be dipped in them and drained, and the japan should be applied with a brush when putting them together, the same as using lead. Red and white lead are also good, but are with more difficulty made air-tight.

If the house is an old one, or has been finished, and you have to test for leaks, take off the meter and cap the bottom of the riser; also unhang the gas-fixtures and remove the brackets, and cap all outlets carefully. Then use ether and locate leaks before tearing up floors or breaking plaster.

The mercury should be made to stand—remain stationary in the glass—if possible, before the work is passed, but a fall of one inch of mercury in an hour would indicate a comparatively tight job.

Occasionally, when a gas-fitter cannot get a job tight, there is a possibility he may cut off the part or floor of the building he cannot get sufficiently tight to suit the inspector's idea of perfection. The inspector can only prove such practice by removing or slacking off a cap here or there about the house if he suspects such an attempt. If no air escapes, then he has the dead end.

WILL BOILING DRINKING-WATER PURIFY IT?

Q. WILL you please tell me if *boiling* well-water will purify it perfectly, so that we can safely drink it, or is it necessary to *filter* it also?

A. The boiling of water will purify it so far as living organisms are concerned, and would probably make harmless water contaminated by cholera or typhoid fever discharges. It would not remove any mineral poisons which might be present in the water, and it is uncertain as to the effect which it would have on certain poisonous substances analagous to strychnine, which are sometimes produced in organic matters by minute organisms. In other words, the destruction of the life of the germs might not necessarily destroy all of their dangerous products.

After boiling, filtration would only be needed in case a cloudiness had been produced in the water, and, as a rule, it is not desirable on account of the risk of contaminating the water by the filter itself, unless this last is quite new.

DIFFERENTIAL RAM FOR TESTING FITTINGS.

ON a recent visit to the workshops of the Boston Water-Works, on Federal Street, our attention was called to an apparatus for testing valves and fittings to high pressures, without having recourse to a hand or force pump, by the pressure of the water in the mains only.

As nearly all valve and fitting makers still use pumps for this purpose, and as the operation is necessarily slow and laborious, we give a sketch of this apparatus and of a yoke modified to hold an ordinary globe-valve, and we have no doubt but that all interested in the testing by hydraulic pressure of fittings, pipes, radiator-bases, sections of radiators or coils, will readily see how it may be adapted to their purpose with saving of cleanliness.

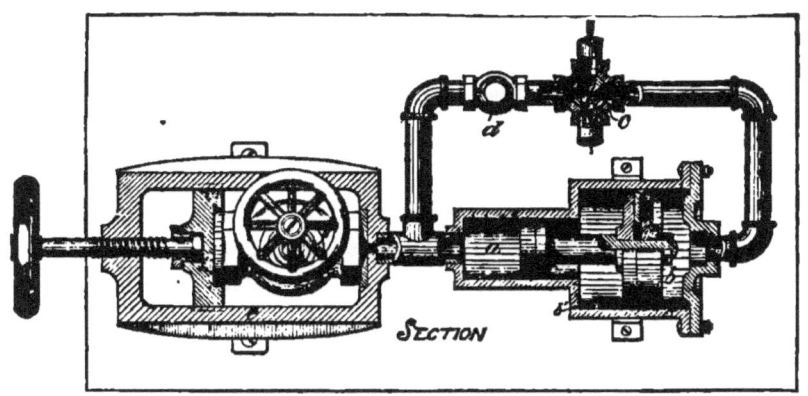

FIGURE 54.

In brief, the apparatus may be called a differential ram. It is composed of a small cylinder and plunger a and a larger cylinder and plunger b—say the larger one being four times the area of the smaller one. These cylinders are connected with the regular water-supply about as shown, a three-way-cock, c, being used for the alternate admission of the water into the cylinders one way or the other.

The fitting to be tested is placed in the yoke e between two elastic washers, and the cock c is turned so that the water passes through the check d, filling the fitting or valve and pushing the piston to the further end of the large cylinder. Then when the cock is reversed, admitting water on the piston b, which is four times the area of the piston a, it follows that should the initial or city pressure be 25 pounds there will be a pressure of 100 pounds in the valve, and this pressure will be maintained constantly, even should there be a small leak, until the plungers have traveled the length of the stroke. The small hole at *b* allows any leakage past the leathers of the pistons to escape.

PERCENTAGE OF ASHES IN COAL.

Q. WILL you inform me what is the proportion of ashes usually to the weight of the coal before the latter is burned? A says it is "fully one-quarter," B says "about one-fifth," while C, who is not an engineer, says that from ten to fifteen per cent. of the actual weight of dry (?) is a fair average for good anthracite coal.

An early reply will be awaited with much interest.

A. You should state the kind of coal under consideration, as in reference-tables on the subject the percentage of *ash* varies greatly with the different kinds of coal. With bituminous coals of well-known kinds —Pittsburg, Cumberland, Welsh, English, and Scotch—it ranges between two and eight per cent., which does not include clinkers or slate.

According to data collected by Chief-Engineer Isherwood, U. S. Navy, Cumberland coal gives about eight to ten per cent. of *refuse*, while anthracite of all kinds gives from fifteen to twenty-five per cent. of refuse according as it is good or bad in quality, and with or without slate.

According to results obtained by other well-known authorities, including Fletcher, Thurston, Emery, and others, the ash and refuse for *all* coals rarely goes under ten per cent., or over twenty per cent. by weight.

According to *twelve* of the published reports of boiler tests made at the Centennial Exhibition, anthracite coal presumably of the best obtainable quality gave 9.9 per cent. as the average.

A and B should bear in mind that to weigh a barrel or two of damp ashes and compare it to the estimated weight of the coal burned for a day is not the proper manner for finding percentage by weight of the ashes to the coal burned. If they dry their ashes or do not wet them, they will find that one-fifth even is too great a percentage of refuse in coal that is purchased as being of good quality.

AUTOMATIC PUMP-GOVERNOR.

A CORRESPONDENT writes:

"SIR: I noticed in the *Scientific American* of April 4, 1885, that a Mr. Frank A. Cushing, of New York City, has patented an overflow-alarm which rings a bell when the tank is pumped full. I will state that I have had such an arrangement on a water-tank for over two years, and it works splendidly, and is in daily operation ever since. It does not ring any bell when the tank is full, but it stops the pumps when the tank is full, and when some water is used out of them the counterbalance opens the valve and the pumps start up again. The valve used is a common 2-inch Coffin valve, and the whole arrangement did not cost me as much as a bell would that any one could hear.

"I will state that we start the pump in the morning, the throttle is never touched until we stop at night, and the tanks are filled from four to ten times a day. The pumps stop and start automatically, and all that is needed is to keep up steam.

"This same arrangement could be used to regulate the draught, and also slow down the pump if the supply should run short.

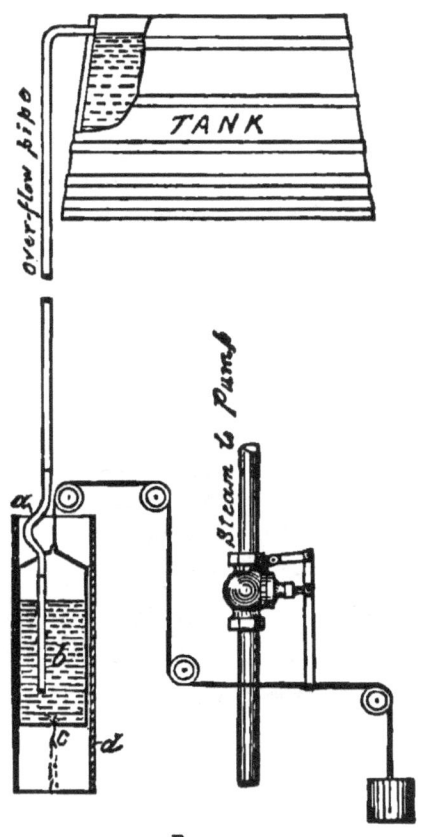

FIGURE 55.

a, Piece of hose on overflow-pipe; *b*, can to receive overflow-water (7 gallons); *c*, small hole in bottom of can; *d*, guide for the can to run up and down in so that the hose will always go into the can. It is nothing more than a common box.

"I will state that there is a small hole in the bottom of the can that lets the water out after it stops the pump, and when the water runs out the counterbalance starts the pumps. I do not think it worth a patent, but I wish other people to use it and see the good of it."

CAST-IRON SAFE FOR STEAM-RADIATORS.

The illustration, Figure 56, shows a cast-iron safe for steam-radiators.

The use of a safe under direct radiators is not particularly new, but their limited use, even in the very best class of buildings, coupled with the advantages that are to be derived from them, especially in apartment-houses, where a leaky radiator or valve on one floor may be the means of spoiling expensive ceilings and furniture on the floor below, induces us to give the accompanying sketch, with a short explanation for the benefit of those who have not seen them in use.

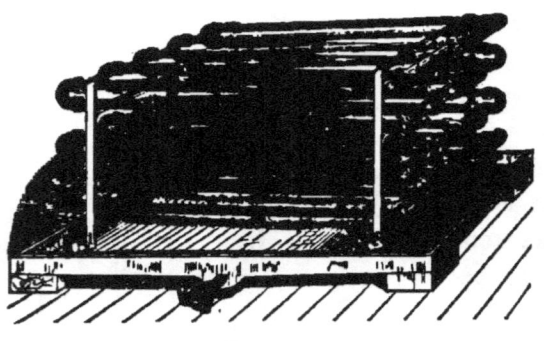

Figure 56.

The safe A is cast with a slight pitch of the bottom to the pipe a, and with sides and ends of about one inch in depth. This is made sufficiently large to go under the heater and the connections which contain the valves, so that all joints that can possibly leak—unless those in the connections between the risers and valves—are protected, the drip from them, if any, falling into the safe.

The pipe a is run backward to the riser recess down which it is carried, and is left open just hanging within sight in the basement, or it is carried to terminate over a sink, which latter is the most appropriate way.

To prevent smell or air from the basement or cellar passing up these pipes into the room, a bend should be used looking up to form a small water-trap of about six inches in depth. These traps are to be filled with water, which will not evaporate if a little ball or clapper valve is placed over the end of the pipe.

The dripping of water from the end of one of these pipes locates the rising line on which the leak occurs and warns the engineer before any serious damage may take place.

We understand the arrangement is not patented, nor is it patentable, and to whom to give the credit for its first use we do not know, but they are now in use in the Dakota apartment-house and other buildings in New York.

METHODS OF GRADUATING RADIATOR-SURFACE ACCORDING TO THE WEATHER.

Q. Is ANY steam-radiator made with which you can regulate the heat? *i. e.*, supposing it takes 60 square feet of surface to warm a given space when the thermometer is at zero outside, and there comes a mild day when half the above surface would be sufficient, is there any radiator made in sections so that you can use one row, two rows, or three rows of pipes, thus regulating the temperature of the room according to the weather?

A. There are several methods now before the public for accomplishing the result you mention. The earliest of these is shown in Figure 57. It has been in occasional use for many years.

It is the plainest form of sectional radiator, being in substance a wall-coil, so arranged that one pipe or any number of the pipes which go to make up the heater may be shut off. In the diagram, $c\ c\ c\ c$, are the usual 1-inch pipes, which are hung on wall-plates, and either run around the corner of the room or "mitered up" on the wall. Instead of the usual headers or branch *tees*, special valve manifolds, $b\ b$, are used. The body is cast-iron, and the nut d and stem and disk d'' are brass, with sometimes a brass seat at e. Thus d'' and e form a valve at the end of the pipe c, the closing of which prevents the entering of steam into the pipe. In like manner and at the same time the

corresponding valve at the other end of the pipe is shut, to keep back
the pressure from the return-pipe. In theory the principle is very

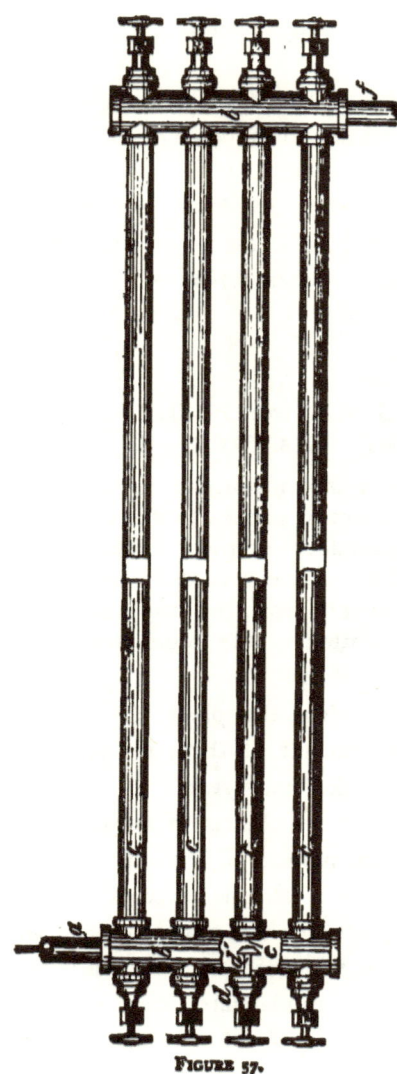

Figure 57.

good, but in practice with steam of even moderately high pressures the valves wear out too rapidly, or they are not properly closed in pairs, or one of a pair may not be closed, in which cases noise is likely to follow. Another objection that has been developed in practice is that should some pipes be shut off during the daytime and the valves leak or pass water by being imperfectly closed, they are apt to freeze during the night or an extreme cold period, if the circulation is not properly established again. In fact, the great number of valves to be attended to makes them impracticable in the hands of any but an expert. With exhaust-steam they may be made to do good work by using one of the valve-manifolds on the inlet end of the coil, having no pressure in the outlet, the condensation being simply allowed to run away.

A modification of the same principle has been applied to vertical radiators by Messrs. Baker, Smith & Co., of New York, for some time, and is shown in Figure 58. It is an ordinary radiator, divided in the base by partitions $p\ p$, so as to separate each row of tubes into a different radiator, as it were ; each radiator or section having its own set of valves—*i. e.*, steam, return, and air valve The pipe d is the steam-supply to a header, a, into which are nipped as many valves, c, as there are sections in the radiator. These valves in turn are connected with the base of the radiator by right and left handed nipples, b. In like manner the return-valve c', header a', and return-pipe d'' complete

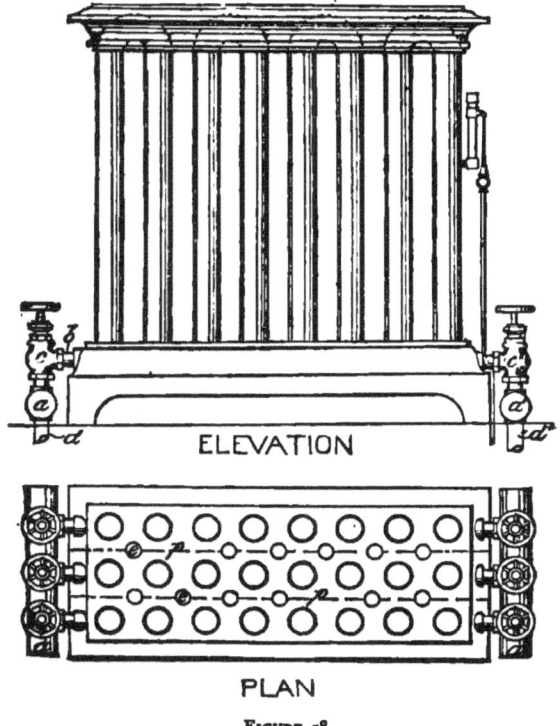

FIGURE 58.

the return-end of the radiator. These heaters are made as wide as six sections, and have small holes, e, through the base, to allow of a somewhat better contact of the air within the pipes than could be had with wide bases if they were not perforated.

Figure 59 shows a different principle for accomplishing similar results. The radiator A may be of any of the approved vertical radiators, or it may be a coil. Steam is carried in the rising-pipe a at a comparatively low pressure, and is admitted to the radiator through a "fractional-valve," *c*, the invention of Mr. Frederick Tudor, of New York, and of which a detailed description and illustration were given on page 616, Volume VIII., of the *Sanitary Engineer*. The valve is arranged for but one revolution of the stem, and is provided with a graduated disk and pointer. In moderate weather the pointer is moved to where the user finds he gets the desired result. With an increase or decrease of temperature the pointer is advanced or withdrawn, admitting more or less steam to the radiator. The maximum opening of the valve is arranged for the greatest surface of the radiator to prevent an overflow or waste. The return-pipe is practically without pressure, and an air-cock may be used either on the radiator or on the return-pipe in the cellar above the water-line for the whole line. To prevent the water from the main return-pipe backing up within the riser *b*, the syphon is introduced as shown, or a check-valve may be used, or both. In cases where a comparatively high pressure is required in the main, the return-riser is carried

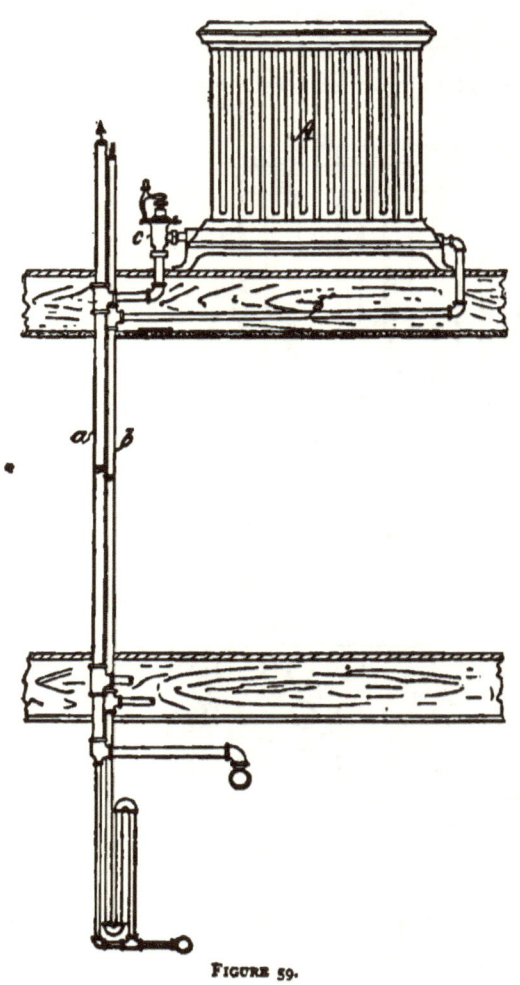

FIGURE 59.

into a separate return-pipe, which runs to a receiver without pressure.

Figure 60 shows the method adopted by Edward E. Gold. It is one of the compound coil radiators inclosed in a sheet-iron case, with a register in the top. Steam is allowed to remain in the radiator most of the time, and the register in the top is used to graduate the amount of warm air passed.

Figure 61 shows a method lately adopted by the Walworth Manufacturing Co. The object is the regulation of the heat of rooms

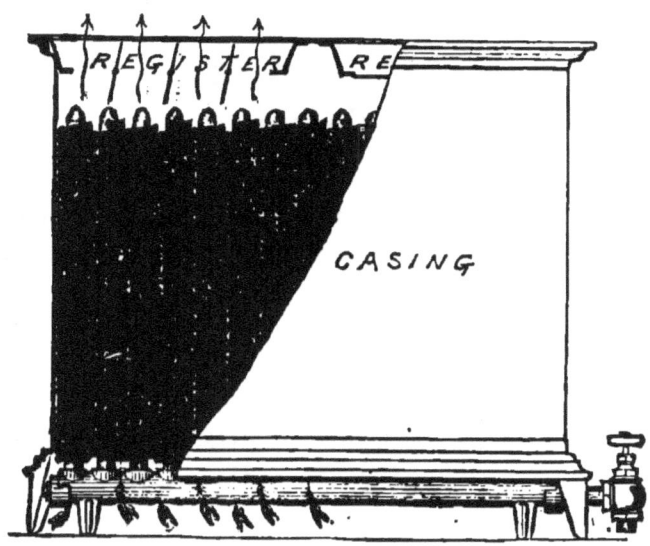

FIGURE 60.

without having recourse to the opening and closing of the radiator-valves, and to place it within the power of the occupants of a room to graduate temperatures.

The figure is a front elevation of the apparatus as it might appear under a mantel-piece as a substitute for a grate, or the radiator, R, may be placed within any recess in a wall or under the inside sill of a window. In front of it is arranged a flexible curtain, C, of metal or other suitable material, the upper end of which is actuated by a spring-roller, so that the screen or curtain may be wound upon it, on the principle of the shade-roller, when the weight of the curtain is lifted by the hand, stopping it at any desired distance from the floor. Fresh air

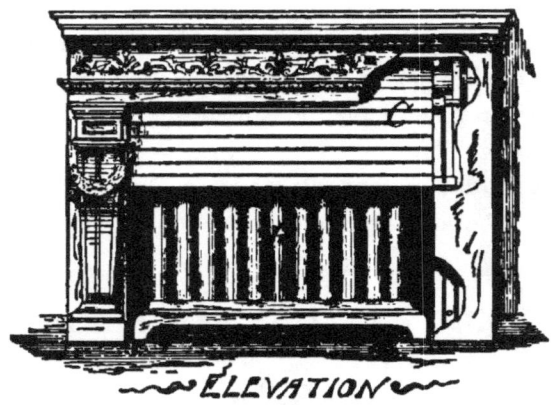

FIGURE 61.

from the outside may be admitted to the inner tubes of the radiator to be warmed and passed into the room, the duct being controlled by a hand-regulated damper or valve.

PREVENTING FALL OF SPRAY FROM STEAM-EXHAUST PIPES.

Q. WHAT do you do in New York City to prevent the water which is carried out of the exhaust-pipes of engines, at the roofs of buildings, from making a shower which will fall into the street? I have been much troubled with spray from this cause, and the contrivances I have used have not accomplished their object.

A. There are several patented apparatus for this purpose before the public. How well they accomplish their object we cannot say, but we hear little or no complaints about a spray nuisance in this vicinity, one reason being the extreme height of most of our buildings in which are engines.

On the roof of the *Tribune* building we find the arrangement shown in Figure 62, which seems to work very well, as it does not appear to wet the roof ; we are informed it is not patented It is a rectangular tank *a*, six feet deep, and three feet by three feet in horizontal

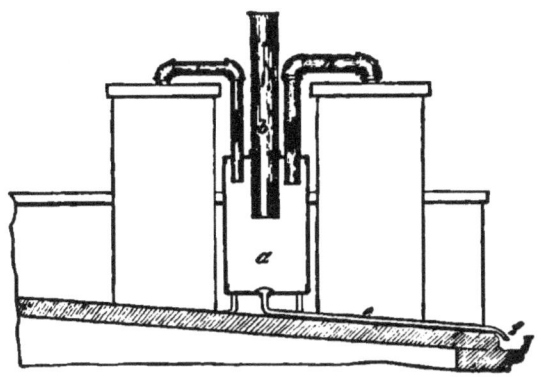

FIGURE 62.

dimensions, within which is fixed a 12-inch pipe, which extends about half-way down. Into this is carried a 6-inch exhaust-pipe *c*, and an 8-inch exhaust, *d.* When all the engines are running only a cloud of very fine vapor issues from the head of *b*, which is soon dissipated in the atmosphere. The pipe *e* conveys the condensed water to the eaves-trough.

EXHAUST-CONDENSER FOR PREVENTING THE FALL OF SPRAY FROM STEAM-EXHAUST PIPES.

THE cut, Figure 63, shows a hood or spray-preventing cap —technically called " exhaust-condenser "—for use on exhaust-pipes from steam-engines to prevent the fall of spray on roof or sidewalk.

The pipe S represents the exhaust-pipe as it comes through the roof or chimney, and the cylinders A and B make up the condenser or trap.

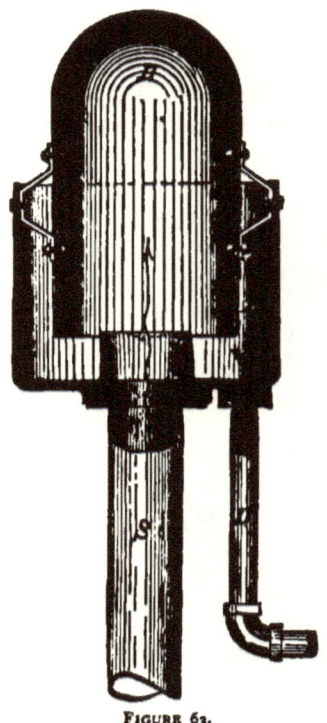

Figure 63.

A indicates a cylinder, open at top and closed at bottom, except at its two apertures *a* and *b*—the former for the entrance of exhaust steam, the latter for the exit of the water of condensation entrapped within the apparatus. From the bottom of the cylinder A a tubular extension or inner nozzle rises, being an extension of the aperture *a*. Above this nozzle, and set partially within the cylinder A, is the inverted cylinder B, secured by angle strap-braces or other braces to the interior of the cylinder A. A waste-water pipe D is tapped into the aperture *b*, and led off to any desirable receptacle, drain, or sewer. When the engine is in operation, the exhaust steam rises through the nozzle *c*, throwing against the inner domed top of the cylinder B any water condensed in the pipes or entrained with the steam. This water is deflected downward and falls back upon the annular bottom of the cylinder A, while the uncondensed vapor escapes out of the trap around the cylinder B through the enlarged annular space between the cylinders B and A, the water discharging itself through the aperture *b* and pipe D. The cylinders B and A may be made of any suitable material, but preferably of cast-iron.

STEAM-HEATING APPARATUS AND "PLENUM" SYSTEM IN THE KALAMAZOO INSANE ASYLUM.

A CORRESPONDENT sends us a detail of the radiator and connecting-pipes, as planned by C. M. Wells, C. E., architect, for the remodeling of the steam-heating apparatus of both the male and female insane asylum buildings at Kalamazoo, Mich.

The buildings were originally planned by Dr. Van Dusen, the medical superintendent at the time, and the heating-apparatus was erected by Joseph Nason, of New York. The system then used was chamber-heating; all the coils for each building were massed at the extreme end of an air-duct, away from the main building, which connected with a passage D, shown in Figure 64, and the air was forced by a fan driven by an engine, sending the warm air through the plenum D and through the flue E to the rooms and halls. In time it was found that though a fan could force sufficient air for the whole building, and there was possibly enough pipe-surface, yet, in windy weather, it was impossible to warm certain halls depending on the direction of the wind, the outside pressure forcing back the warm air. The management, then, as a matter of experiment, added supplementary box-coils placed under the entrance to the flues E, which were found to need them most. The experiment proved satisfactory, and gradually the engineer, under the direction

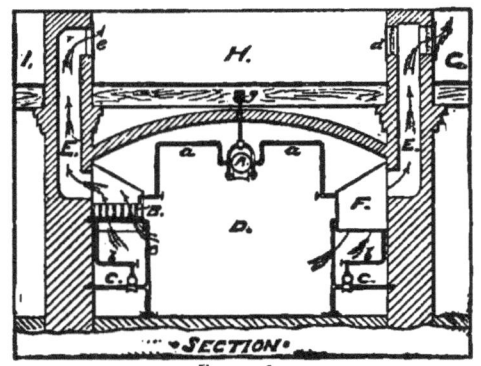

FIGURE 64.

of the medical superintendent, removed the original coils, distributed them throughout the plenum, and boxed them with wooden casings. In this condition (necessarily a patch-work) the apparatus was used, the fan being still retained until the spring of 1881, when the Legislature voted money, at the request of Dr. Palmer, the present medical superintendent, and the commissioner, for the entire remodeling of the apparatus on the same principle with regard to the disposition of the heating-surface, and in respect to piping adopted the low-pressure gravity system, still retaining the fan, though not always having to use it, as each flue has its separate heater with the new system.

148   STEAM-HEATING AND STEAM-FITTING PROBLEMS.

The diagram is a cross-section of the passages under the buildings, with a total length of about 2,200 feet. A is the main steam-pipe hung on expansion hangers, which are adjusted by a nut in the bracket G ; *a a*, the steam connections to radiators, are purposely elbowed up and down as shown, to compensate for the expansion of the main ; B is a radiator of the cast-iron extended-surface pattern (much liked by the engineers), and *b b* are the return-water connections to the main returns C C. The object of having two main return pipes is to prevent the too frequent crossing of the passage with small pipes ; F is a galvanized-iron coil casing or hood, arranged to be taken off for cleaning or repairs, and is open at the bottom, as shown by the arrows.

With this method it has been no trouble to get a uniform heat throughout the buildings, each flue doing its full duty independent of the others.

HEATING AND VENTILATING A PRISON.

The accompanying diagrams, Figures 65 and 66, illustrate the principles of the warming and ventilating of the New York State Reformatory at Elmira, N. Y.

Figure 65 is the detail of the radiators, which present some novel features.

Figure 66 is a cross-section of one of the wings of the building, showing a "cell block," of which there are four, with the arrangement of the air-passages and the course of the air from the time it enters fresh at the window *a* until it escapes through the exhaust-chamber *c*.

In prisons the problem of efficiently warming and ventilating is complicated by the necessity of providing for the safe keeping of the inmates, and it requires methods not ordinarily met with and of unusual strength of construction.

The heaters are round vertical-tube radiators set under the windows, with openings in the centres of the bases. In corresponding openings in the stone flags are set strong cast-iron pipes *t*, with flanges built into the masonry. These pipes extend up through the openings in the bases of the radiators which they fit closely, connecting the fresh-air ducts with the radiators and preventing water (when washing the floors) from entering the ducts. The upper end of the pipe *t* is also fitted with a strong plate, *f*, which acts as a baffle to the entering air, forcing it between the tubes of the heater, and also as a cover to the pipe, so that nothing of large size can be passed through it.

The number of concentric rows of tubes in the radiators is four. The two outer rows are separated from the inner ones by a galvanized sheet-iron partition or septum *r*, the object being to divide the inside rows from the outer ones so as to make part of each radiator practically an indirect heater, the air from the duct only coming in contact with the inner rows, while the outer rows warm the air already within the halls and give direct radiation.

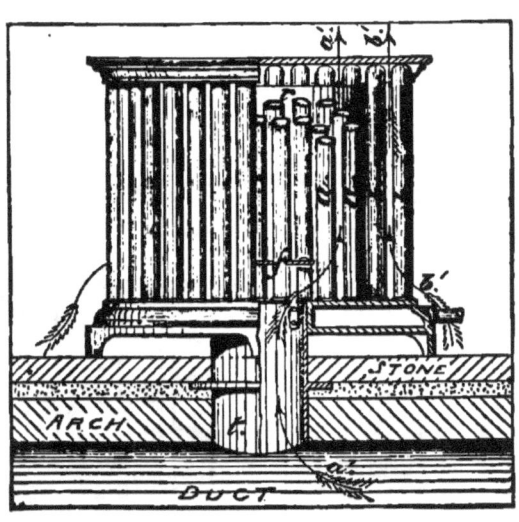

FIGURE 65.

The galvanized sheet-iron partition is simply a sheet of No. 20 iron, bent and riveted at the edge and slipped into place, and it may be used with any round radiator, making what is called a "direct-indirect-radiator."

It is claimed that the short shaft or flue formed by this simple device accelerates the movement of the air to a very considerable degree, and prevents the possibility of its being drawn down into the duct, by preventing a reversal of the movement of the air in the towers.

For radiators arranged in this way, it is said that marble tops are not as good as cast-iron fret-work, as the former hinder the easy ascent of the current of air; and we believe that for all purposes the

150   STEAM-HEATING AND STEAM-FITTING PROBLEMS.

marble top does not give quite as high an efficiency for the same reason.

There are 500 cells, each of which have two 4x4-inch flues, one from near the ceiling and the other from a cast-iron niche near the floor The one near the ceiling is fitted with a heavy cast-iron frame

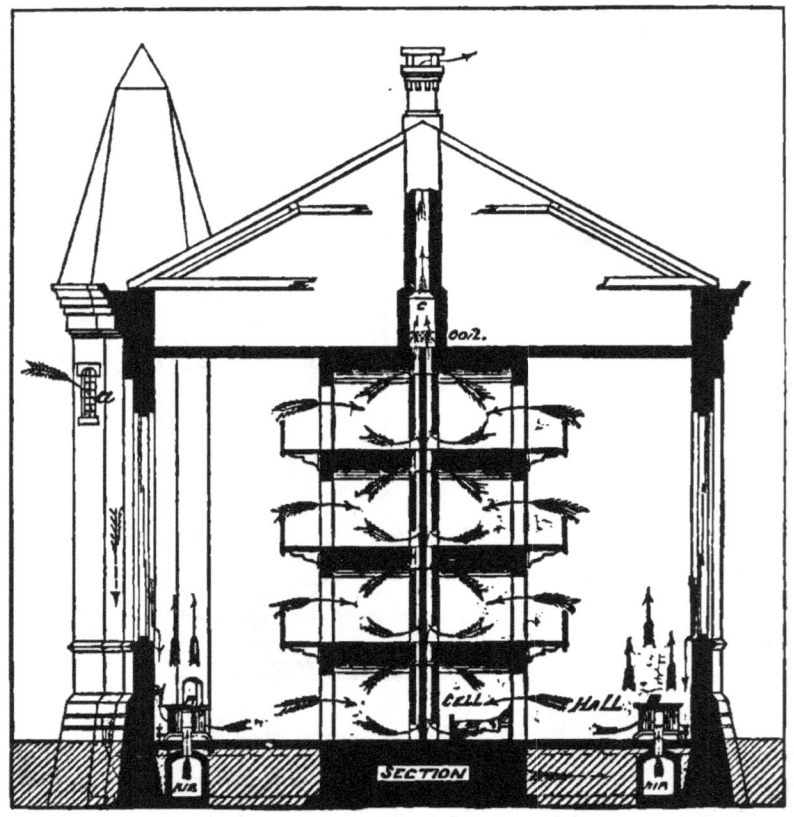

FIGURE 66.

built into the walls, while the lower one connects with the top of the "night-bucket" niche. The flues are separated their whole length, each terminating in the exhaust-chamber c as shown, and there is no means of closing them.

The exhaust-chambers extend the whole length of the blocks of cells, so that the flues are perfectly straight, a person in the chamber being able to see the light in the cells.

The steam-coils within the exhaust-chambers are of $1\frac{1}{4}$-inch pipes and extend over the upper ends of all the flues.

The air-ducts extend all around the wings near the outer walls as shown, and communicate with the fresh-air shafts or towers, each tower having a separate section of the duct.

The course of the fresh air is in at the window a and down through the tower, thence through the air-ducts to the radiators, through which it passes to the halls, from which it is drawn into the cells by the action of the independent flues, and thence passes out through the aspirator.

The coils in the aspirator or exhaust-chamber are not connected with the regular heating system of steam-pipes, but with a special system provided for them, and with the valves in the boiler-room, so that they can be under the control of the engineer without his entering the buildings; this also admits of using the exhaust-chamber and coils during the summer and when steam is not otherwise required, so causing a rapid movement of air at all seasons.

We are informed that when the air is moved sufficiently often to give each cell one-half a cubic foot of air per second, the difference of temperature between the upper galleries and the flag is only five degrees in the coldest weather.

The system of steam-pipes is "gravity-return," and the pressure is increased or decreased to suit the weather.

The boilers are ordinary horizontal tubular, five in number, 16 feet long by 4 feet in diameter. Two, and sometimes three, are used on the heating apparatus, and one for cooking, drying, etc., while one is always in reserve.

We are indebted to Z. R. Brockway, General Superintendent of the prison, under whose supervision the institution was built, for the drawing of the building.

AMOUNT OF HEAT DUE TO CONDENSATION OF WATER.

Q. WHAT I have been informed is a late and novel method for determining the amount of heat imparted by steam in a steam-heating apparatus consists in measuring or weighing the water condensed in the apparatus, and therefrom estimating the amount of heat due to condensation. The successive steps are: First, supplying steam at a known pressure and temperature; second, condensing the steam in the apparatus; and third, measuring or weighing the water. Is not this the only method, and is it properly stated?

A. When steam is used for warming air which is unconfined, having free access to the glass of the windows and to other cold surfaces, as well as changing constantly through the ventilators, there is no means of measuring the heat which has been given off except by weighing the water which has been formed within the pipes.

If the condensed steam (hot water) is received in a vessel which is open to atmosphere and weighed at a constant temperature—say 200° Fah.—the units of heat given off will be 978.6, whether the steam had one pound pressure above atmosphere or 40 pounds, and is equivalent to the warming of 939 cubic feet of air 50 degrees.

To make it necessary to consider pressures and temperatures, the water would have to be kept under the same high pressure and at the same temperature as the steam was; in other words, if you cooled a pound of steam from 40 pounds pressure to water at the *same heat and pressure*, there would be only 893 *units* of heat given off—equivalent to warming 857 cubic feet of air 50 degrees.

The first method is practicable, and can be done by drawing the water into a bucket, weighing it, and taking its temperature with a thermometer. The second method cannot be carried out without some especial contrivance for weighing the water within the pipes, at the same time keeping the temperature of the water constant while waiting for it to accumulate.

EXPANSION-JOINTS.

THE sketch, Figure 67, illustrates a form of joint designed to obviate the tendency to "telescope," which is one of the objections to most expansion-pipes. The flange A is for connection to stop-cock (stop-valve on top of dome), and B is the steam-pipe leading to

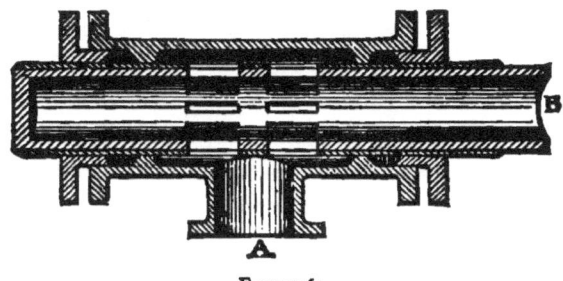

FIGURE 67.

adjacent boilers. Stops may be fitted to the latter outside of each gland to prevent any attempt to slide out, in case of emergency, but otherwise they are not required. In fact, I am of opinion that the ordinary socket pipe is safe enough itself; and I have never yet seen a boiler-seating which was not amply capable of sustaining all the side pressure caused by the steam forcing a surface equal in area to the pipe.—C. R., in *Mechanical World*.

[C. R.'s expansion-joint is a form not common to steam-fitters and users, not being a regular article of stock with any of the houses which make a specialty of such goods.—ED.]

RESETTING A HOUSE-HEATING BOILER—A POSSIBLE SAVING OF FUEL—BEST METHOD OF FIRING.

Q. I HAVE a horizontal boiler set in my house for warming purposes. It is very expensive to run in the matter of fuel, and the brick-work is so dilapidated and cracked that I am advised to reset it. Will resetting and stopping the cracks save much of the fuel, and if I reset it will I be any better off at the end of another year? The bricklayer who did the work the first time says I will not, and that all boiler-walls crack.

A. If your foundations and boiler-walls have sunk and cracked badly there is reason to assume that the walls rest on sand, mud, or made land, and resetting will probably greatly improve the boiler. In any case excavate and put in about ten inches of good concrete over the whole of the bottom. Then, when set, pave the whole with hard-burned bricks set on edge, and on this begin your walls. This will prevent excessive cracks for the future, but will not prevent one crack in the side walls near the middle, which nearly always appears in the outer sides, and is due to the greater expansion of the fire-bricks and inside linings of the furnace. The concrete makes a unit of the bottom of sufficient strength to support the boiler and walls. Many think it unnecessary to do this in wet sand, and cite the cases of other foundations when they reach damp sand. But their view will not hold good for boiler or furnace work, as the heat drys the sand and makes a quicksand of it. Another reason for putting in a solid concrete bottom is to cut off moisture.

It may also be that your grate is too large, and that you waste fuel in this manner. Eight pounds of coal per square foot of grate per hour will give best average economy, but presumably with slow combustion, such as you must have with an automatically regulated boiler, you had better proportion the grate for from five to six pounds of coal per hour. If you then consider *thirty* pounds of coal per hour to each 1,000 superficial feet of radiators in the building good practice, you will be able to determine the size of grate you should use. This would call for six square feet of grate, and should the boiler be thirty-six inches in diameter with 1,000 feet of surface, a 30″x34″ or a 30″x30″ grate will be required. Not knowing whether to charge the waste to the size of grate or to poor settings, we give the above dimensions so that too much will not be expected of the new settings, should you decide to make the improvement.

A properly set horizontal tubular boiler is not a wasteful one, as few classes of boilers do better.

HOW TO FIND THE WATER-LINE OF A BOILER—POSITION OF TRY-COCKS.

Q. Some time ago I bought one of Baldwin's works on steam-heating, and though it gives valuable information, it does not tell me how to get the exact water-lines in various makes and shapes of boilers or what proportion to allow for steam and what for water. Where should the highest try-cock be, and also the lowest when put directly into the boiler, and what proportion do you allow between each cock in proportion to the size of the boiler?

A. When steam is drawn from a boiler with a regular water-feed, the steam being used for supplying engines or other purposes, and when the water is not returned, all that is required is a safe water-line; and by a "safe" water-line is meant one sufficiently above the crown-sheets or tubes to give time for interruptions in the management of the feed-water.

In a heating-apparatus where the water returns by gravity the case is different. With such apparatus there should be water enough above the crown-sheets or tubes to fill all pipes or radiators with steam at the highest pressure ever likely to be carried before the water-level falls to a dangerous water-line; and not only this, but a little more, as at times of first starting a gravity-apparatus the water does not always begin to "come back" or return properly when the pipes are first filled with steam. The water, of course, may be provided for by letting water into an apparatus which acts this way, but after the apparatus is in *train* it has to be let out again and will cause trouble and danger, as it will some day result in the blow-off cock being forgotten for a few minutes.

To find the amount of water necessary to have in a boiler above the safe-level for a gravity apparatus, calculate the cubic contents of all pipes and radiators and the space above the water-line in the boiler in cubic feet and divide it by 1,061, 718, 614, 510, and 435 for 10, 20, 30, 40, and 50 pounds of steam respectively, and the answer will be in cubic feet of water. By doubling the amount thus obtained you will be safe with any gravity apparatus which will work ordinarily well in returning its water.

The distance between the try-cocks is an arbitrary matter, and no positive rule can be laid down for it, two or three inches being common.

LOW-PRESSURE HOT-WATER SYSTEM FOR HEATING BUILDINGS IN ENGLAND.

For dwelling-houses, a "register-stove" or "range" in each room is the favorite means of heating in England; and, while this plan heats a *small* room tolerably well, it is evidently a very imperfect mode of heating a large room, while for large public buildings and greenhouses it is simply useless.

A comparatively short time ago the hot-air system was largely used in England, but now it is almost entirely superseded by the "low-pressure hot-water" system.

This method of heating is effected by means of a boiler, in which the heat is generated, and a system of pipes laid around the building, which conveys the heat from the boiler, and distributes it in the various rooms where it is required.

The usual temperature of low-pressure hot-water pipes is from 180° to 200° Fah., and it is nearly impossible for them ever to obtain a higher temperature than 212°.

It can be said in favor of low-pressure hot-water that there is *no danger* from excess of pressure or through the carelessness of unskilled attendants.

Like all other inventions, however, its first application was simple and limited, but gradually its principles have become so well known in England that there is no building too large or too complicated to be warmed on this plan. To understand the system correctly, and thus enable it to be successfully applied, some knowledge is necessary of the course of heated water circulating in pipes, and of the principles on which the successful working of an apparatus depends. Without this knowledge errors will be made in the construction of an apparatus, which will cause its failure. In fact, this has often been the case in past years, and in some instances has caused apparatus to be condemned as worthless, when in reality the ignorance of the hot-water engineer alone was to blame.

Two theories have been advanced to account for the circulation of water when heated through a system of pipes: one by Tredgold, in his work on "Heating by Steam," and another by Hood, in his work on "Warming Buildings."

Hood's theory, which is manifestly the correct one, is briefly stated as follows:

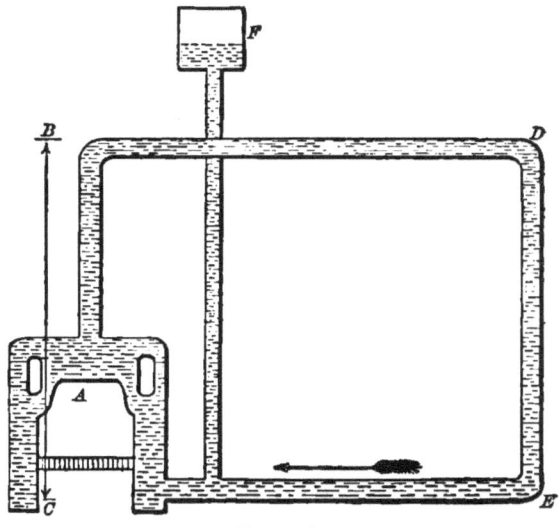

FIGURE 68.

On the application of heat to water *at* or *above* 39° Fah., it increases in volume and decreases in density; therefore, a column of water at 120° Fah. is *heavier* than another column of the same height but heated to 180°. This principle operates in a hot-water apparatus in the following manner:

In Figure 68, A represents the boiler. On the application of heat the water contained in the boiler becomes heated, and the heated particles rise to the point B in the flow-pipe. The column B C—*i. e.*, the water contained in the *boiler* and *flow*-pipe—expands, and becomes lighter, bulk for bulk, than that which is contained in the return or descending pipe D E. This causes a somewhat greater pressure at E than at C, and in consequence of this extra pressure at E the water is forced along the bottom pipe into the boiler, as shown by the arrow. Just as the colder water enters the boiler, an equal amount of *hot* water leaves it by the flow-pipe. In this way a constant circulation is kept up. The water leaves the boiler heated. It is constantly giving out warmth as it flows through the pipes, and is, therefore, colder as it gets further away from the boiler—the flow-pipes always being warmer than the returns. So long as heat is applied to the boiler this difference of temperature of necessity exists, and a constant circulation is the result. F is a feed-cistern regulated by a ball-cock, and connected to the return-pipe near the boiler.

Every hot-water apparatus may be said to consist of two parts— viz., (1) a boiler, or that part of the apparatus in which the heat is generated and imparted to the water, and (2) a system of pipes (with the necessary air-vents), which may be called the heat-radiating surface, or that part of the apparatus which conveys the heat from the boiler and distributes it in the various parts of a building, as may be required.

In erecting an apparatus, the following are vital points, which must always be kept in view : The boiler should be the *lowest* part, and the pipes running from the boiler must gradually ascend, without any dips or depressions under doorways or otherwise, until they reach their furthest extremity. All ascending pipes are flow-pipes, and the descending pipes are returns. The water *always* flows from the highest part of the boiler, and re-enters at the lowest part, or as nearly so as possible.

The apparatus shown by the figure is of the simplest form, with only one circulation, but, as previously stated, it is capable of being applied in an equally simple, but more expensive manner, so as to be sufficient to warm the largest public building, warehouse, factory, or range of greenhouses.

A very important part of the apparatus, and that which makes it a success economically, is, of course, the boiler ; or, in other words, the means for utilizing the fuel to best advantage. A large variety of designs have been introduced into the market at various times, both of the "independent" types and of those requiring brick-work.

Independent boilers are very useful in many cases, but can only be profitably employed where the length of pipe to be heated is small. Generally speaking, it is best to have a boiler set in brick-work. In these latter boilers the fuel is used to much better advantage than is the case with "independent boilers," because the flames and heated gases are carried by means of flues all round the exterior of the boiler after the fire-box and flues have absorbed all the heat possible during the passage of the gases of combustion. By this means the greater part of the heat generated in the furnace is utilized instead of a large portion of it being allowed to escape up the chimney, as is the case with many independent boilers, without doing the least good, or having the least effect in the way of heating water. In addition to this, a considerable amount of heat is radiated from the external surface of an independent boiler, unless it be covered with a non-conducting substance, such as silicate, and, as the boiler itself is usually fixed in the cellar, where this heat is seldom or never required, it represents so much heat and a proportionate amount of fuel wasted.

The qualifications of a good boiler are simplicity in design, durability and economy in consumption of fuel, combined with efficiency, and a reasonable price. By always remembering these very important requisites, it is easy to see that the *cheapest* boilers sold, *if not carefully selected*, will probably prove the most expensive in the long run. Every boiler should be made of wrought-iron, not cast-metal, as cast-iron boilers are liable to crack without a moment's notice, and that too, perhaps, when the apparatus is most required. Many boilers are now made fulfilling all the necessary conditions already named, and may be described generally as boilers of the "saddle" form, with one or more return-flues, both internal and external. "Saddle boilers" are *always* preferable to upright boilers, as horizontal heating-surface is so much more effective than heating-surface which is vertically disposed.

An English Hot-Water Heater.

[COMMENT ON THE PRECEDING BY THE EDITOR.]

Though the principles explained are thoroughly understood by our American heating engineers, yet to many readers they will be instructive and interesting—notably, to a large number of men in the plumbing business who are unfamiliar with the principles of hot-water circulation in connection with domestic range-boilers.

The question of absolute safety cannot be gainsaid when considered in connection with the apparatus shown and described, and for horticultural purposes it is conceded by all to be the most practicable method of heating.

In the United States it is used only to a limited extent in residences and single offices or suites in the *direct-radiation form—i e.*, where the coils are in the rooms—for the reason that the space occupied by the water-coils is largely in excess of that occupied by direct steam-radiators, especially on account of the coils having to occupy a horizontal position, and because they must be larger in diameter and greater in amount of surface. A great many of the Government buildings throughout the country are warmed by indirect low-pressure hot-water apparatus, in which coils of 3-inch cast-iron pipes form the heating-stacks, through which the fresh entering air has to pass on its way to the rooms and apartments. Many private residences in New York and vicinity are also warmed by indirect hot-water methods, some of which are low pressure, though a few are on the "Perkins system," or a system somewhat akin to it, in which the temperatures are kept at a maximum of about 260° Fah.

By the direct method, such as our correspondent describes, many of our city houses, with fronts of twenty-five feet or thereabouts, in blocks, with windows and cooling-surfaces in front and rear only, could be warmed to good advantage. Of course, we should have to tolerate horizontal pipes of large dimensions across the front and back walls under the windows, and we are satisfied that direct radiation under such conditions is better than indirect radiation from a "fire-place heater" down-stairs, which warms about two rooms overhead, and than some furnaces which take air from a lower room or basement.

With regard to the comparative merits of cast and wrought iron boilers for hot water, there is room for considerable difference of opinion, and, assuming all things are the same except the natures of the metals, it is reasonable to suppose that like boilers will do like duty, though the cast-iron boiler will be the most easily broken; although, on the other hand, makers of cast-iron boilers claim that, with certain well-known forms, the result of gradual development, the question of breaking is reduced to a minimum, and that they are able to make forms that cannot be made in wrought-iron. However, this latter is a purely practical question, which cannot be decided in one general answer.

The *Sanitary Engineer* offices and printing-office are warmed by a low-pressure hot-water apparatus, simply because we found it best suited to our own particular situation, and in adopting it we did not mean to imply that we considered it the best for all places. Hot-water heating is largely in use in Canada and the maritime provinces for residences and buildings of limited size, but for large buildings steam is now being more generally used.

STEAM-HEATING APPARATUS IN THE MANHATTAN COMPANY'S AND MERCHANTS' BANK BUILDING.

The illustrations show the plans of the steam-heating apparatus, boilers, pipes, pumps, etc., of the Manhattan Company's and Merchants' Bank Building, 40 and 42 Wall Street, New York.

From Figure 69 a conception can be obtained of the operation of a "graduated system" of steam-warming in a large building.

The term "graduated system" comes from the use of a "graduating-valve" on the radiators.

It is well known that with most systems of steam-warming in general use, the steam must be turned on full to a radiator; in other words, the valves at both ends of the heater must be wide open when heat is required, or the result will be loud noise in the radiators or coils. This, in only moderately cold weather, will make the room too warm, provided the radiator is proportioned for cold winter weather, and there is nothing left for the occupant but to turn on and off the steam, as often as he finds the room uncomfortable, either in one direction or the other. This is obviated to some extent by having several radiators in a room, one or more of which can be in use or closed, as desired, or by having one radiator with two or more sections, each section controlled by a set of valves, or by inclosing the radiator in a case with a register in its top. In the present case the object is to have the radiator—and the extent of its surface warmed—within the control of the individual who sits near it, and to leave it within his power to graduate the surface so warmed by as fine subdivisions as he pleases, by the manipulation of a single valve, a description of which will be given hereafter.

In brief, the system of heating is as follows: Steam is taken from the boilers B (Figures 69 and 70) at any pressure—fifteen to twenty pounds being usual—to the regulating-valve R V. There it is reduced automatically to a *constant* pressure of *two* pounds per square inch; thence it is conveyed for distribution through the 8-inch low-pressure main steam-pipe (L P M S) to *two* 6-inch rising lines (M S R), up which it passes to the main distributing-pipe, eight inches in diameter, at the ceiling of the eighth or top story, M D P in the diagram, Figure 69, and in the eighth-floor plan, Figure 71. From the main at the eighth story it is distributed downward through pipes, which conventionally are called *risers*, and which, though they distribute downward in this case, are still known as "main distributing rising-pipes," marked M D R P, one of which is shown in the diagram, and all of which are shown on the eighth-story plan as the diagonal dotted pipes.

These pipes are principally three inches in diameter where they leave the main on the upper story, and terminate two inches in diameter at the basement-floor, below which they are 1¼ inches as relief-pipes only to where they join the main return-pipes—M R in diagram and cellar plan. Parallel to these pipes, but starting at the floor of the eighth story and running downward, are the "main return-risers." These pipes are entirely without pressure, other than atmosphere, and are sealed against the pressure from the main return-pipes by the water-seal S, which is five feet six inches in depth, and sufficient to resist

MISCELLANEOUS.

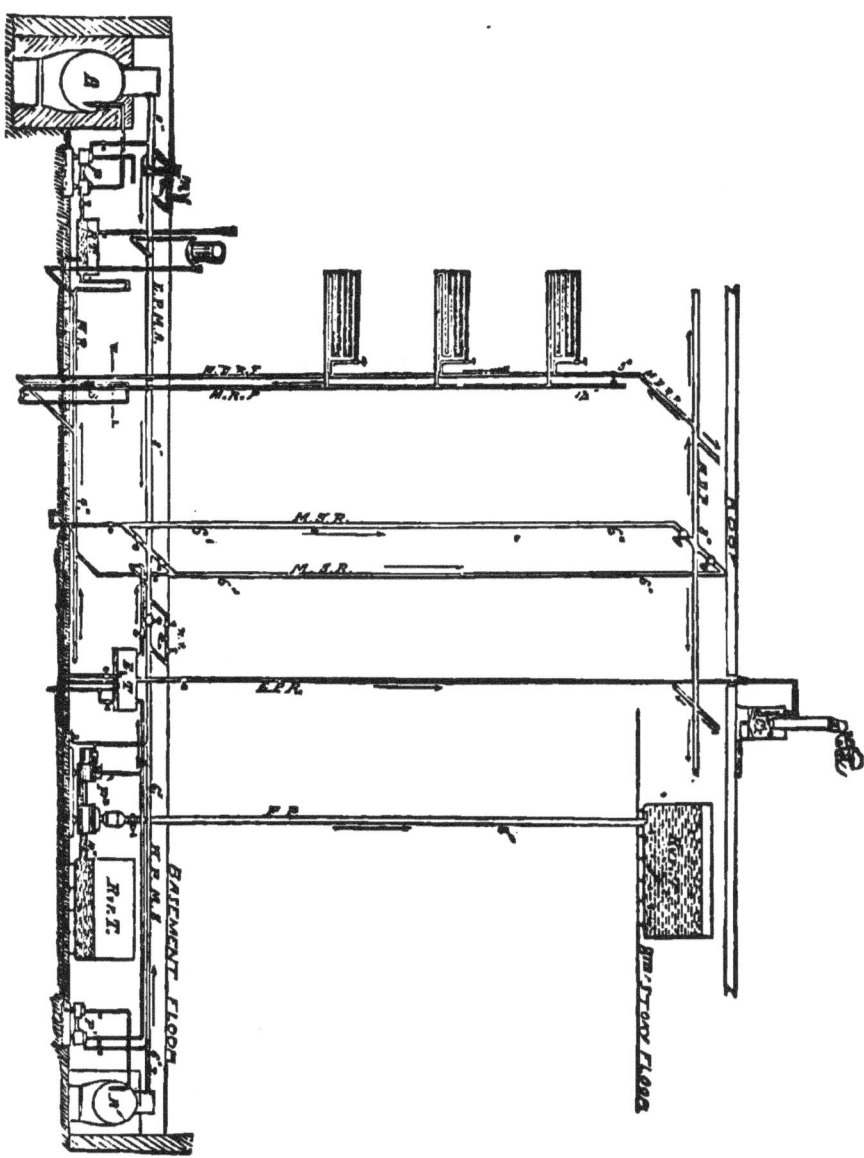

FIGURE 69.

two pounds water-pressure from the main returns, even when the water is very hot. These pipes are practically open to the atmosphere above the water-seal. A *tee* is put into them below the basement floor, but above the water-level—W L in diagram. To this tee is attached an automatic air-valve, which is set to remain open so that air may pass in or out, for the purpose of preventing " air-binding " in the coils between the steam as it enters through the graduating-valves and the water in the seals. It may, for the sake of convenience, be called a breathing-hole, and would be allowed to remain open, and without an automatically or thermostatically regulated valve, were it not for the possibility of the regulating-valve (R V) failing for a moment, and allowing a high pressure to pass into the "returns" and raise the water. But as an extra precaution against that, a check-valve has been introduced into each water-seal at the bottom, as shown, opening inward, to let the water of condensation which runs down the riser return-pipe enter the main return-pipe, a slight accumulation of head of water in the riser-return furnishing the power.

The course of the steam, then, is from the pipe M D R P through the coils to the pipe M R P, the latter of which it is never intended to reach as steam, the condensed water only running into it, and falling by gravity into the water-line within it.

At the upper end of the coils is the *graduating-valve*, Figure 74, in the position shown; it is a finely-regulated valve, with small opening about one-quarter of an inch in diameter, with a long, tapered spindle filling the hole. On the spindle where it passes through the nut of the valve is a coarse thread, one revolution of which will withdraw the spindle from the hole. To the handle of the spindle is attached a stop-arm, which engages stop-pins on a graduated circle. The full opening of the valve has been found by experiment, and is regulated to pass only the amount of steam that the radiator is capable of condensing in an atmosphere of 70° Fah., leaving fifteen per cent. of the coils which is always to remain cold to prevent an overflow of the steam into the return-pipe. Any quantity of steam between the maximum and minimum is then secured by bringing the stop-arm to the desired graduation on the circle.

To provide for changes of temperature of the outside current, and consequently greater condensation within the building, there is a full thirty per cent. of coil-surface which it is not necessary to bring into use, except in a cold room when first turning on the steam, or in extremely cold weather.

The two 6-inch main rising lines are relieved in the usual way by 1½-inch pipes, and every down line is a relief for the main at the top of the house.

The main return-pipe terminates in a receiving-tank, R T. Before entering the tank it rises to a height just equal to the height of the water-seals (S'). The object of this is to seal the lower ends of the relief-pipes from the (*down*) steam-risers M D R P, the pressure within the risers being only *two* pounds, minus loss by condensation and friction. This prevents the "blowing through" of the main returns to the tank, and keeps all pipes sealed. From the receiving-tank the water of condensation is pumped into the boilers by the pump P.

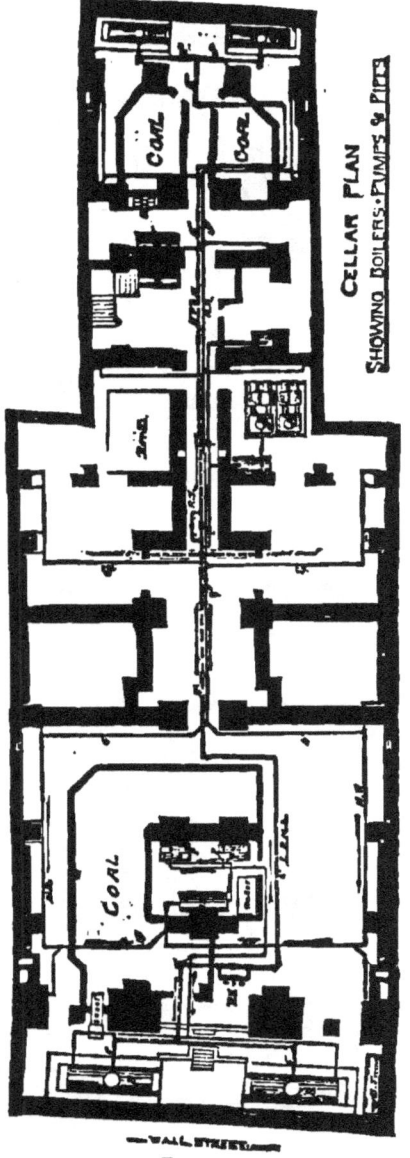

FIGURE 70.

The object of selecting *two* pounds as the normal pressure in the apparatus is three-fold: (1) With the ordinary height of cellar (eight feet), water-seals deep enough for a high pressure could not be obtained and have any factor for safety; (2) the exhaust-steam from the hydraulic pumping-engines enters the heating-pipes in winter and cold weather, and is there all condensed; and (3) graduated valves are more accurate and less liable to "cut," with a difference of about two pounds pressure between the steam supply-pipes and the coils, or, in other words, between *two* pounds and atmosphere, than with higher pressures.

The pumps P^2 in Figures 69 and 70 are duplex compound pumps, with high-pressure steam-cylinders 12 inches in diameter, and low-pressure cylinders 18½ inches in diameter, with 12-inch water-plungers, forcing against a head of water of 58 pounds in repose and 62 pounds when in action.

The pressure carried in the boilers B' (high-pressure boilers) is 60 pounds. The terminal pressure in the low-pressure cylinders of the pump is about 17 pounds, and the back-pressure 2½ pounds. From these pumps the exhaust

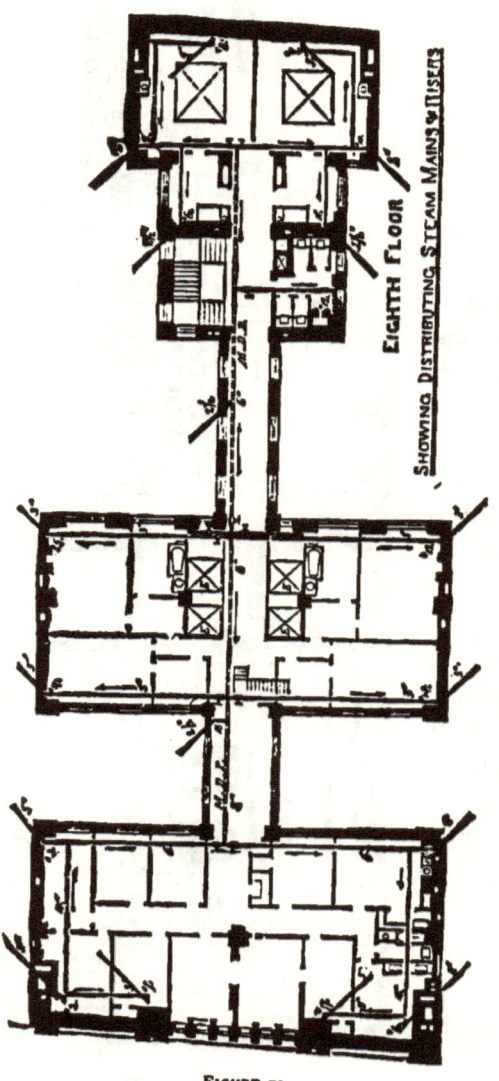

FIGURE 71.

steam goes to a tank, F T. Here the water of condensation from the cylinders of the pumps, or water held in suspension, and grease are separated from the steam, which is then passed to the heating-pipes, as shown, in the winter time, or to the roof-condenser C in the summer.

As the exhaust-steam from the high-pressure boilers and pumps, after passing through the heating system, and being condensed therein, must reach the receiving-tank as condensed water, the pump P' for the high-pressure boilers also takes suction from this tank. This prevents the wasting of this water, and is the means of supplying the high-pressure boilers with condensed water, which is passed through the heater tank H T on its way to the boilers.

The hall-radiators are not on the graduating principle, but take steam from the lower main in the old method. This is shown at the right of the diagram in a circular radiator.

The main steam-pipe from the high-pressure boilers is also arranged to give steam to the heating system through a "pass-by" and regulating valve, shown near the centre of the diagram. The object of this is to provide a means for a temporary supply from the high-pressure boilers should the low-pressure one be out of order, or to give steam in the mornings and evenings of fall and spring, when it is desirable not to run the low-pressure boilers should the exhaust-supply not be sufficient.

A peculiar feature of the warming of this building is the skylight heating. Over the principal counters, in the centre of each bank, is a domed skylight, 24'x16' at the base, and shown in section through its greatest measurement in Figure 72.

As the air cooled by this glass and iron structure would certainly fall upon the heads and shoulders of the officers and clerks who were underneath it, and whose desks were arranged with special reference to the light, it became necessary to devise some method for the warming of the air near where it was cooled, and to prevent, if possible, its falling.

To accomplish this and retain the architectural effect unmarred, the architect provided in the main rafters of the roof for the introduction of a 3-inch pipe, P, Figure 72, in such a manner as to simulate a rail—or, perhaps, more properly speaking, a cord—of the rafters of the dome. This passes, as shown, through the roseate, and is bronzed to correspond with the iron-work of the dome, and appears to be an integral part of the structure.

Above this, at C, further provisions for warming are made. The cast-iron shell-cornice which covers the structural iron-work and beams is removed from the latter three inches, and a coil of six $1\frac{1}{4}$-inch steam-pipes is concealed in the space thus formed. At the bottom an opening or slot one inch in width is made, which extends all around, but is so arranged, by one member of the cornice so overlapping another, as not to be apparent from below. The top of this space is

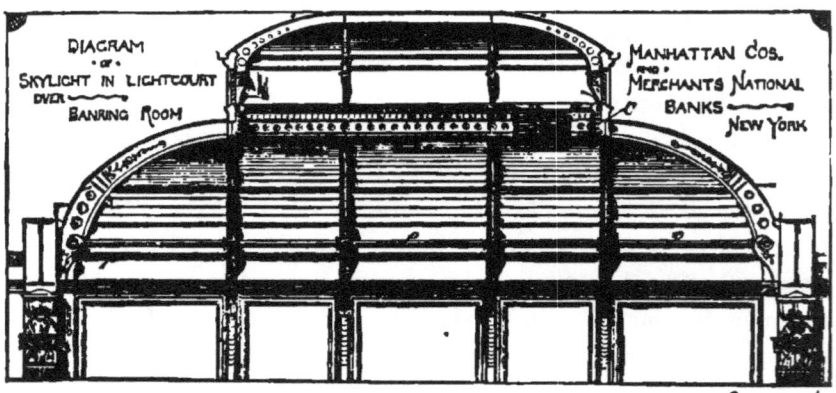

FIGURE 72.

also open and a current of air is established which comes in contact with the pipes and is delivered warm in front of the windows in the upper part of the dome.

These windows are pivoted to swing outward at the bottom, and in ordinary cold weather are kept a little open. This produces ventilation and a gentle outward current of air in a position where it ordinarily will come in. Over the directors' rooms are skylights similarly warmed.

Figure 73 (to the left) shows skylights at the rear of the banking-rooms. In this case a 3-inch pipe, P, is also attached to the skylight-rafters by a special bracket, and the identity of the heating-pipe is lost in the finish. At the top, at C, is a coil of eight 1½-inch pipes, eighteen feet long, which also assists to warm the air in this vicinity as well as to warm the glass and surrounding iron-work and marble by direct radiation. The method of warming under the windows on the Wall Street front is shown at the right (Figure 73). No air is taken in, but the warm air from the coil is made to pass up in front of the plate-glass windows and neutralize the down current from the glass, which would otherwise fall on the heads of the clerks.

The coils in skylights are not on the "graduated system," but take their steam at full pressure, with their stop-valves in the cellar under the engineer's control.

Figure 74 shows the graduating-valve made for this building. In many respects it is an ordinary angle-valve, of strong and very accurate make. It is so designed that one revolution of the stem and handle gives the maximum opening of the valve. The "pitch" of the *thread*

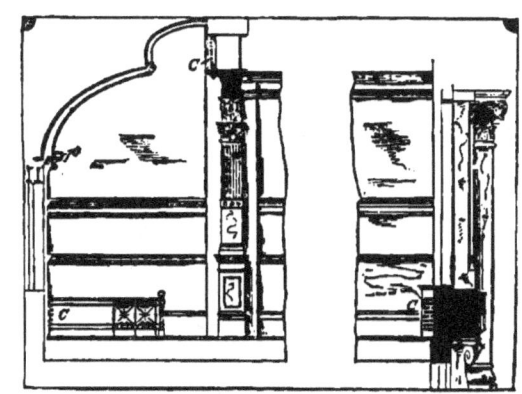

FIGURE 73.

on the stem is six threads to the inch, and the hole at the bottom is gauged to pass about the greatest quantity of steam that the heater can condense. The long taper of the disk has a three-fold object: the first is to allow of drilling the hole through the bottom of the seat any size, to suit any size heater, and still form a perfect valve; the second is to admit of considerable movement of the handle and a backward motion without a rapid increase of annular space between the disk and the seat; and the third is to prevent a singing noise, by gradually allowing the steam to expand through the annular space, which space increases in area as it advances up the cone of the disk. Two stop-pins are placed in the graduated circle, one at O and one at the maximum opening of the valve, which opening is determined by experiment, and the second pin put in. Should this pin be removed by any person and the valve-stem turned beyond it, the fixed size of the hole at the bottom, which is approximately correct,

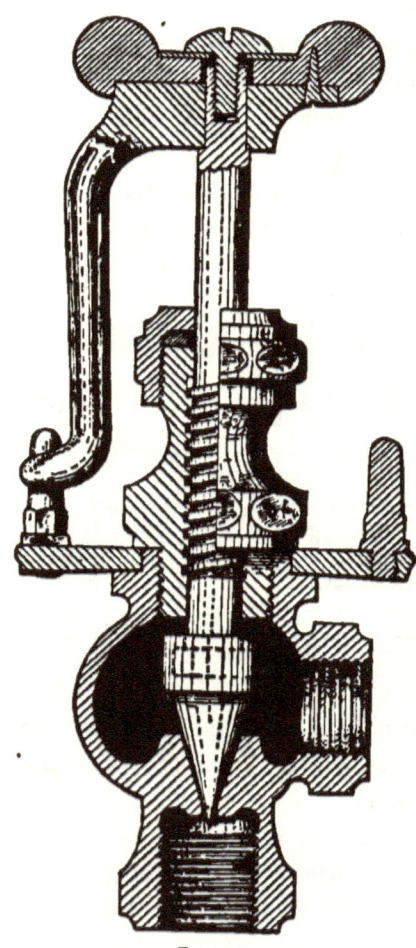

FIGURE 74.

will prevent any serious overflow of steam. The round hole, seven-thirty-seconds of an inch in diameter, at the bottom of one of these valves was found to pass thirty-six pounds and eleven and one-half ounces of steam in one hour, with the pressure in the pipe two and one-half pounds by the steam-gauge, the lower end of the pipe-coil into which it flowed and condensed being open to atmosphere.

Figure 75 shows the tank used in this building for the separation of grease from the exhaust steam that is turned into the heating-mains. The steam enters through the 6-inch pipe *a* at the left, and escapes through the pipe *b* at the right to the heating-main. The large sectional area of the tank between the pipes results in a comparatively slow motion of the steam, giving the particles of water, grease, etc., an opportunity to fall into the water *w* in the bottom of the tank, allowing comparatively dry steam to escape. As the quantity of water in the tank will increase if it is not regularly drawn away, the contrivance *d*, *c*, and *e* is provided. The water overflows through the bend *c* into the trap *e*, which is of sufficient depth to withstand the pressure within the tank (from two to two and one-half pounds). The short pipe is carried from the bend down into the water for one-half its depth, as this is supposed to be the best point at which to draw off the water, so as not to take oil from the surface or dirt or heavy fats from the bottom of the water, and to prevent the stopping of the pipe by accidental accumulations.

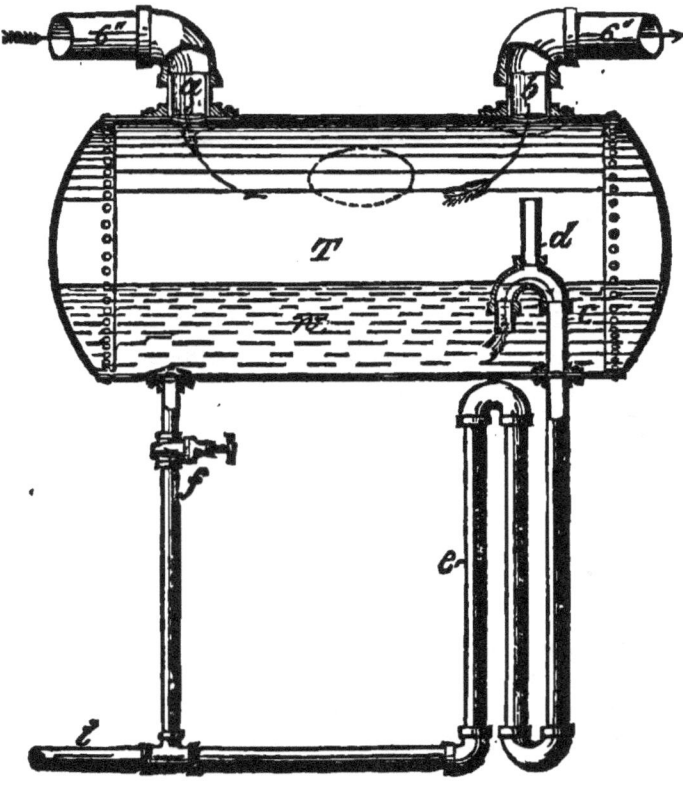

FIGURE 75.

The pipe d is put in the top of the bend to prevent the syphoning of the water from the tank, and is carried sufficiently high to be above the highest level of water in the tank.

The pipe f is a draw-off pipe, and the pipe j runs to the blow-off tank.

The pipe *e* of the trap should have a valve in it at any desirable point, the object of it being to "choke" the pipe to such an extent as to prevent the fluctuations of pressure of exhaust steam starting the water out of the seal.

The architect of the building was Mr. W. Wheeler Smith, who was assisted in the engineering department by Mr. W. J. Baldwin, M. E. The contractors were Messrs. Bates & Johnson, all of New York.

THE BOILERS IN THE MANHATTAN COMPANY'S AND MERCHANTS' BANK BUILDING.

FIGURES 76, 77, and 78 show the boiler and boiler-setting of one of four similar boilers lately set in the Manhattan Co.'s and Merchants' Bank Building, 40 and 42 Wall Street, New York. The special points about these boilers are : The quality of iron used, the method of bracing both head-sheets and domes, and the making of a number of small holes (3-inch) through the shell of the boiler under the dome instead of cutting out a large piece, as is usually done.

To the engineer or steam-fitter the drawings convey a clear conception of the manner of construction, but as the specification from which these boilers were built was drawn with a view to give instructions to the builders as well as to secure the desired results to the owners, we quote from it such parts as will be of interest:

" To furnish and put in place and set according to these specifications and the plans, two low-pressure multi-tubular boilers, each to be sixteen feet in length from head-sheet to head-sheet, by sixty-six inches in diameter, and each to contain 102 3-inch outside diameter charcoal-iron lap-welded boiler-tubes, not less than No. 12 wire-gauge in thickness, and

"*Domes.*—Each boiler to have a dome on the top of the shell thirty-six inches in diameter by thirty-six inches high, measuring from the centre of the top of dome. At the front end of each boiler the shell is to extend before the head-sheet twenty-one inches, making the extreme length of a boiler seventeen feet and nine inches.

"*Manholes.*—Each boiler will have a manhole, with its plate, etc., in the top of the shell back of the dome, the manholes to be 15x11 inches, and be set with their longest diameters across the shells.

"*Handholes.*—Each boiler will have two handholes, with their plates, etc., one placed in each head in the position shown, and of the sizes shown.

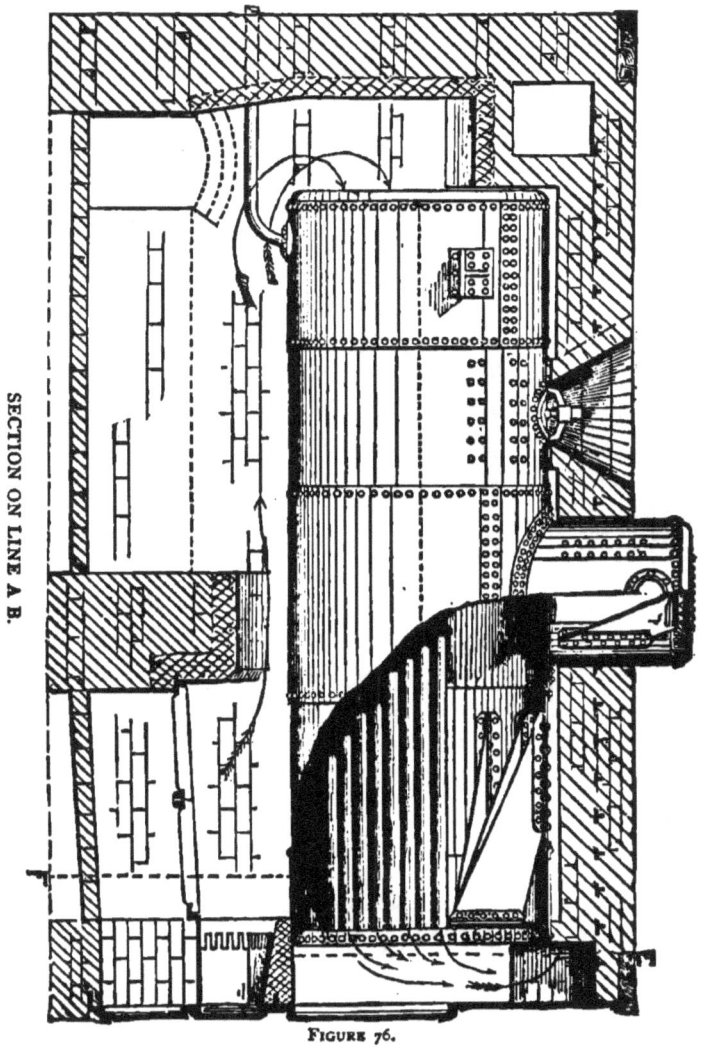

FIGURE 76.

"*Iron.*—The shells and head-sheets and domes of said boilers to be made throughout of "extra flange fire-box iron," bearing the stamp and name of some reputable manufacturers of boiler-plates, that will be satisfactory to the architect or his representative, and every plate or part of a plate used in the construction of these boilers will be so marked in one or more places —thus, "Extra Flange Fire-Box Iron," together with the usual stamp (initials or otherwise) used to designate the makers.

"*Samples.*—If in the opinion of the architect samples of the different plates should be required for testing, the contractor must furnish them from each or any number of the plates, and must prepare them for the testing-machine in the usual manner, and must pay the cost of testing them and for the report of the expert who performs the tests. And should the plates prove to be a lower quality than that above stated and called for, the contractor is bound to furnish other plates until the standard of quality is reached and secured.

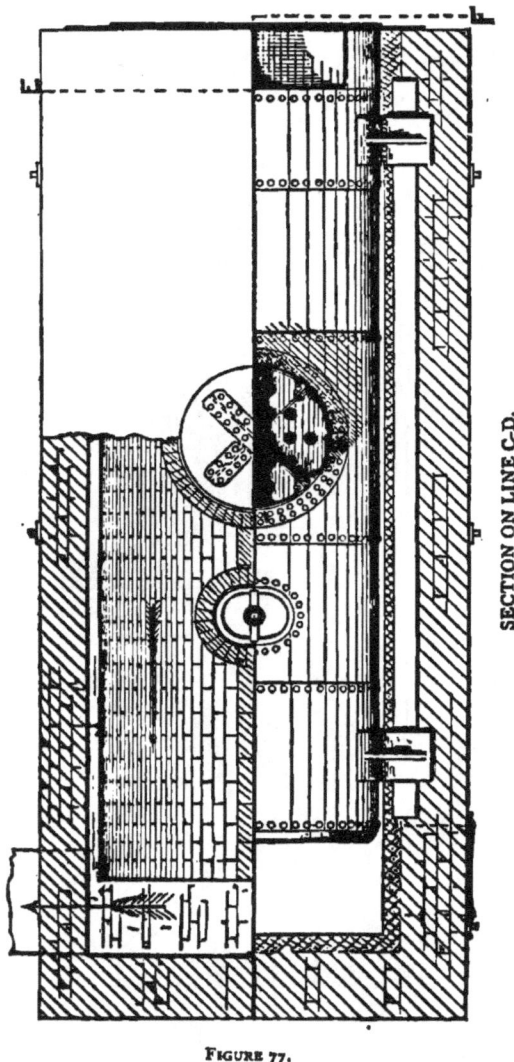

FIGURE 77.

"*Thickness.* — The shells or cylindrical courses of these boilers are to be full three-eighths of an inch in thickness, and to have but one longitudinal seam in each course, the seam to be in the upper quarter segment of each course.

"*Heads.*—The head-sheets of said boilers to be made of ½-inch thick iron of the quality before stated, the angles of whose flanges will be bent to a radius not less than 2½ inches, and which must not show flaw or blemish after being turned, and which must be carefully annealed before being punched.

"*Domes.*—The domes of said boilers to be made of the quality of iron before stated, and to be five-sixteenths of an inch in thickness, with ½-inch thick heads, and flanged with not less than 2-inch radius in the angles of the flanges, and braced in the manner shown, for which a detailed drawing will be furnished.

"*Bracing.*— The head-sheets of said boilers will be braced in the manner known as "flat gusset-bracing," each head having five braces, a detailed drawing of which will be furnished.

"*Seams.*—All longitudinal seams in the boilers must be double riveted, and the rivets so spaced from centre to centre, and also be of such size and diameter as to give seventy per cent. of the strength of the *solid plates* of the boilers.

"The side seams in the dome must be double riveted, and the seam uniting the dome to the shell must be double riveted.

"The seams that unite the courses to each other and the head seams are to be single-riveted seams (presumably ¾-inch rivets, 2½ inches pitch, is the most desirable, while in the longitudinal seams two rows of ¾-inch rivets, 2¾ inches pitch, and placed in the manner known as "square riveting," will give the result required above). The double riveting in the dome-flange must be also square riveting, the flange being

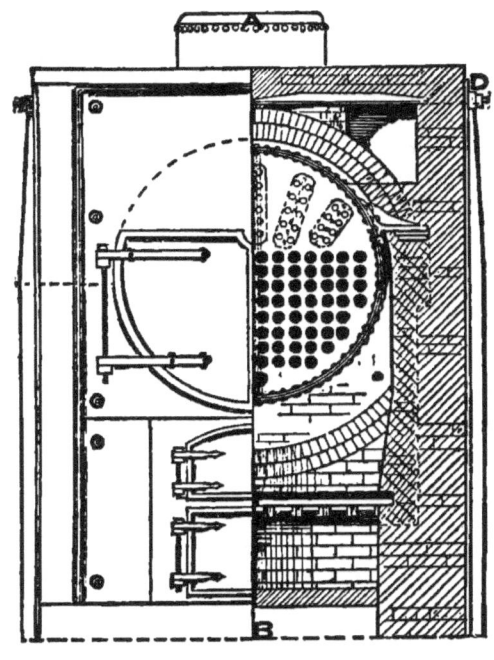

FIGURE 78.

HALF ELEVATION AND SECTION ON LINE E-F.

turned with sufficient depth to place them in this manner, and the "pitch" of the rivets may be reduced to 2½ inches, so as to get the strength of the inner row of rivets *in their length* more nearly equal to the strength of the metal in the dome, but not close enough to each other to reduce the strength of the remaining parts of the shell between the rivet-holes to below seventy per cent. of the solid plates.

"*Rivets.*—All rivets to be of a grade and quality of iron acceptable to the architect, to whose satisfaction they are to be tested as to their tensile and shearing strength by the contractor, and at his expense, if called on to do so.

"*Riveting.*—Machine-driven rivets are to be used in all seams of the boilers in which it is possible to drive them, and hand-driven rivets will be only accepted or allowed in places where machine rivets cannot be driven, or are impracticable, and will then only be accepted when they are driven in a manner to the satisfaction of the architect or his inspector.

"*Calking.*—All the seams of the boilers are to be properly chipped and calked, the calking to be what is known as "Connery's calking." No split calking will be allowed.

"A furrow from the chipping-chisels or calking-tools, or from any cause, in any plate or on any part of the boilers, or the breaking or splitting of a hole from the use of the drift-pin or otherwise, or a crack or flaw or burn in any plate or head, will be deemed a sufficient cause for the total rejection of the boiler or boilers in which any of the above is found to exist.

"*Drift-Pin.*—The use of a drift-pin in the construction of these boilers or in the construction of any one of them, either in the "fitting-up" of them or in the preparation of the holes for the rivets, will be the cause of the rejection of all the boilers.

"*Holes under the Domes.*—The shells of said boilers underneath the domes are not to be cut out in one large piece, but to be a number of 3-inch holes aggregating not less than 100 square inches, and laid out in the manner required, for which a detailed drawing will be furnished. Special attention must be paid to this, for should the shell be cut out in a manner unsatisfactory it will be considered a cause for rejection.

"*Tubes.*—The boiler-tubes are to be set as shown in the detail drawings, and are to have 1¼ inches between the tubes in the vertical rows, and one inch in the horizontal row, and be otherwise set as shown.

"*Expanding.*—They are to be expanded with Dudgeon expanders, with the extension beyond the heads of uniform length, and not broken or ragged by a chipping-chisel, but cut with some proper tool at the end last expanded, or, better still, to be of the proper length before they are inserted.

"*Lugs.*—Each boiler is to be furnished with four (and no more) cast-iron lugs or brackets for the purpose of supporting the same in the brick-work. These lugs are to project from the sides of the boiler 12 inches, and are to have a projection below the plane of the bracket, as shown, with three ⅞-inch rivets in this lower projection and the usual complement above.

"*Manholes.*—The manhole castings are to be extra heavy, and subject to the approval of the architect, who reserves the right to have them made from designs of his own if they are not acceptable or sufficiently heavy in his opinion.

"*Handholes.*—The handholes to be furnished with plates, bolts, and guard, the guard on rear handhole to be "turtled-back," with a projection around the nut to protect it from the fire.

"*Flanges.*—The pipe-flanges on the domes to be of the size marked on drawings and in the positions marked, and to be riveted to the domes and chipped and calked on the inside, and set true with the axis of the boilers.

"*Flanges on Shells.*—A 2½-inch flange to be placed at the rear end of the rear course of the shell at the bottom side, and to be set true, and riveted and chipped and calked. A 2-inch pipe-hole to be drilled and tapped into the rear head and 1½-inch in the front head where shown.

"Two of the boilers are of the size stated above, and are intended only for heating at a low pressure. The other two are each 11 feet long by 48 inches in diameter, with seventy-two 2½-inch tubes, and are for high-pressure steam to run the compound pumping-engines."

STEAM-HEATING APPARATUS IN THE MUTUAL LIFE INSURANCE COMPANY'S BUILDING ON BROADWAY.

THE accompanying sketches illustrate the principle involved in the new heating apparatus lately put into the Mutual Life Insurance Company's *old* building, 140 Broadway, corner of Liberty Street, by the Steam-Heating Division of the New York Steam Company.

Figure 79 is a diagram illustrating the principle involved, and Figure 80 is a plan of that part of the ground floor of the building given up to engineering purposes.

There are three horizontal boilers, each 54 inches in diameter by 16 feet long, containing seventy-four 3-inch tubes, from which steam is taken for heating and power purposes. The steam for power supplies two pumping-engines for hydraulic-elevator purposes, and two ordinary engines for power and electric-lighting purposes. The pumps are one duplex compound non-condensing Worthington engine, with 12-inch high-pressure steam-cylinders, 18½-inch low-pressure cylinders, and 10-inch water-plungers. The other is a simple duplex, 18-inch steam-cylinders by 10-inch water-plungers, with 10-inch stroke. These pumps when in use run intermittently, their duty being to keep the pressure tanks of the elevator service supplied with water, and are governed and controlled automatically by the height of the water in the pressure tanks. The intermittent pumping supply to the tanks produces an irregular action in the use of steam, and consequently gives an intermittent supply of exhaust steam. In the summer time this exhaust steam, of course, is allowed to escape by the roof-pipe, but in times when heat is required on the building it is allowed to pass into the heating-pipes and there be condensed. The supply of exhaust steam from the other engines is also allowed to pass into the heating-pipes, though in their case the flow is constant.

The object to be accomplished now is to condense all the exhaust steam in the heating-pipes so as to save its heat; but at the same time it becomes necessary to provide means of keeping a constant supply of say two pounds per square inch in the heating-pipes, whether the exhaust steam is sufficient to do it or not at all times, and to make up any deficiency there may be, either from lack of volume or the pumping-engines remaining idle for any time.

To accomplish this, direct connections are made between the boilers and the heating-mains, the connections being controlled automatically.

By the aid of the diagram, Figure 79, this may be made plain; it is best to trace the steam from the boilers through the two different courses by which it reaches the heating-pipes.

To this end steam is taken through the front steam-connections of the domes and supplied to the "high-pressure steam-pipe," shown in

MISCELLANEOUS.   179

both figures. Thence it is supplied to the pumping-engines, E P, returning through the "exhaust pipe" to the tank E T—*i. e.*, exhaust tank—in which any grease is separated from it in the manner hereafter described, and in which steam is stored to a certain extent ; the tank being a reservoir and equalizing-pressure tank to prevent a sudden

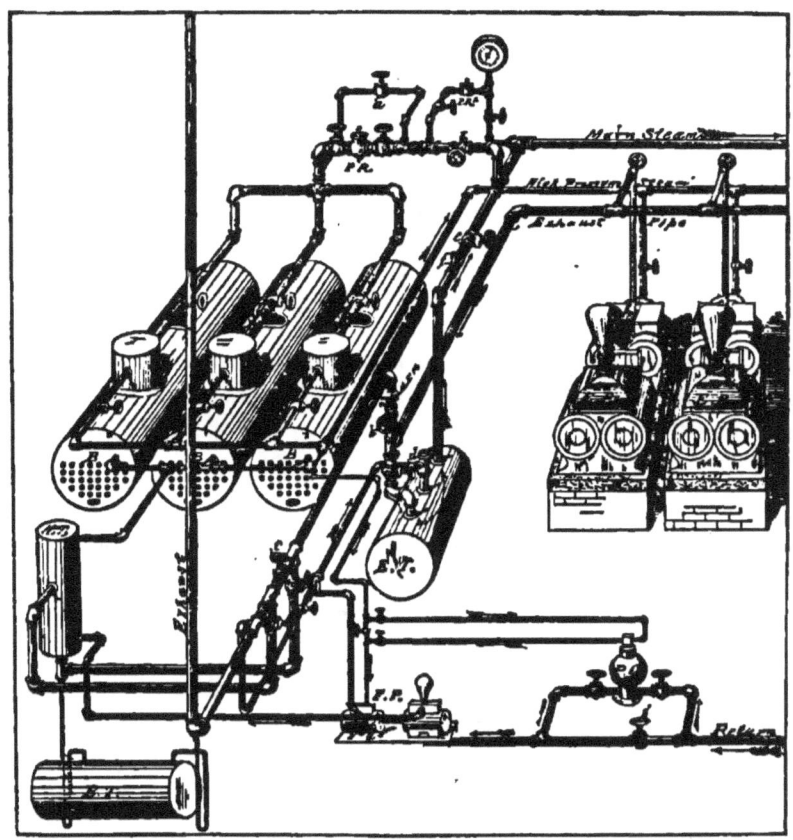

FIGURE 79.

or material increase of pressure when the pumps suddenly start up, and to have a supply to draw from when they are idle for a few moments.

The pressure from this tank passes into the heating-pipes through the swinging check-valve *f* and the stop-valve C, thence to the "main steam-pipe" as shown.

The other method of a constant supply, whether the pumps and engines are running or not, is from the connections on the back of the domes to the pressure-regulating valve P R, and thence through the valve b to the "main steam-pipe." When reducing through a large regulating-valve from high-pressure steam to very low pressures noise sometimes follows, and frequently irregularities of pressure. To prevent this, the secondary pressure-regulator P R² is introduced, and the valve b is closed. Then the steam is reduced, say from eighty to twenty pounds through the first valve, and from twenty to two through the second one, insuring a constant and more regular pressure in the steam-mains. Should the supply from the pumps increase the pressure in the tank (E T) to above two pounds, the regulating-valve admits no more live steam; but should the pressure in the tank fall much below two pounds—say one-quarter of a pound—the regulating-valves open and the pressure is maintained.

The tank E T, as mentioned above, is used for separating grease carried from the engines or pumps, and is sometimes called a "skimming-tank" for that reason. To accomplish this it is kept about one-third full of water. The steam blows down in one 6-inch pipe, and escapes through the other, as shown by the arrows, Figure 79. In passing through the large tank the velocity becomes almost *nil*, and separation goes on by gravity as well as by being blown against the surface of the water. To draw the constantly increasing water from the tank, a bent syphon is introduced into the tank, the top of the bend being at the water-line of the tank, and the open end extending into the water about six inches. This causes the overflow of any excess of water from the tank, and draws it at a distance below where floating oil will be, and sufficiently above the bottom of the tank to prevent sediment from running off in the same way. The water drawn away in this manner is sufficiently free from oil to return it to the receiving-tank, but the usual method is to run it into the sewer to prevent the possibility of getting oil into the boilers this way. To prevent the syphoning of the water from the tank through the bent pipe described above, a pipe is carried from the highest part of the bend to a distance, say one foot above the water. This lets the air or steam pressure into the pipe and destroys

MISCELLANEOUS.

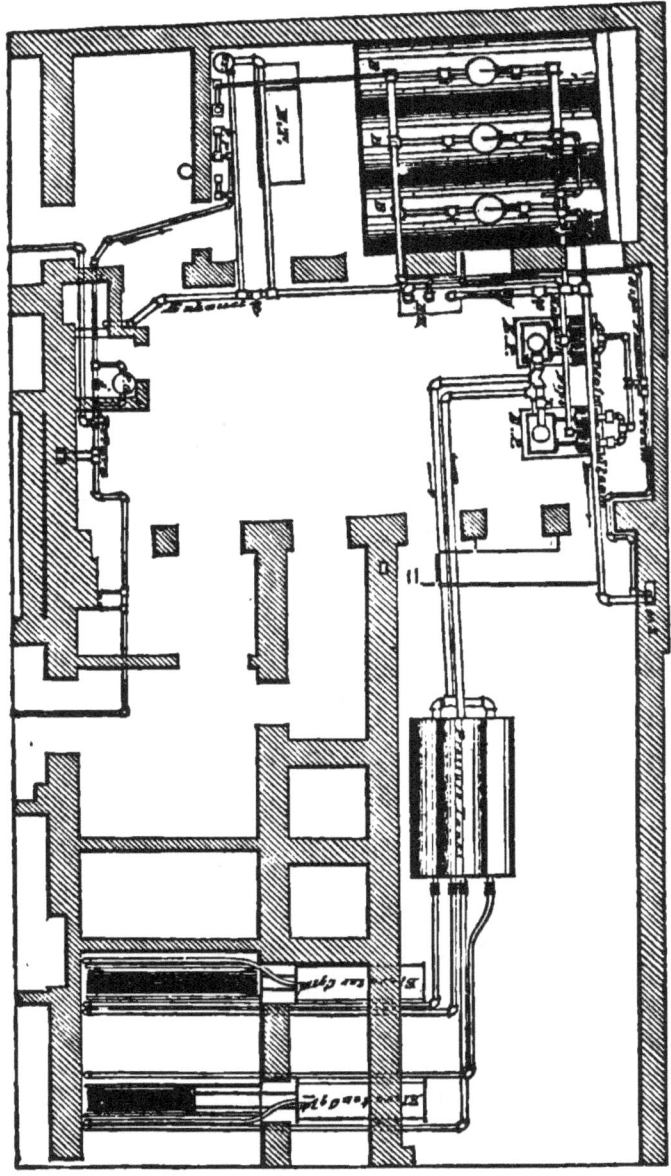

FIGURE 80.

the power to syphon. At the bottom of the tank is a blow-off pipe to be used for drawing off the water and grease or oil.

The check-valve f prevents the pressure from the pipes passing back into the tank should the regulating-valves get out of order, and in such case, should the pumps or engines be running, the back-pressure valve B P V opens by an excess of one pound pressure, and allows the exhaust steam to escape to the roof for the time being.

The small pump F P returns the water of condensation to the boiler, accomplishing this automatically by the aid of the pump-governor P G. The tank H T is a feed-water heater, and the one B T is a "blow-off tank." The same pipes can be traced in plan in Figure 80, and this figure also shows the elevator service, with the "Hinckley" pressure-tanks.

The principle of piping involved in the heating apparatus is to carry all the steam to mains at the ceiling of the upper story, and run around the building, feeding downward. Where the first "riser" is taken from the steam-main, the "return-pipe" commences in the cellar and runs in the same way as the steam-pipe, though fully 100 feet below it, increasing in size as the main steam-pipe decreases, and ending by making a circuit of the building, throwing steam and water always in the same direction.

The designer of the apparatus was H. M. Smith, Division Engineer of the New York Steam Company.

THE SETTING OF BOILERS.

FIGURES 81, 82, 83, and 84 show the setting of the *Tribune* Building boilers. It shows the difficulties attending boiler-setting in localities where superficial area of ground is worth as much as it is in lower New York, and the trouble of securing enough of it at any price is very great.

In this case, in Spruce Street (the street being narrow), it was necessary to go considerably beyond the curb-line to even get two boilers side by side outside the line of the columns of that part of the building; but to complicate matters there was no way of reaching the chimneys except underground; and in consequence of being limited in length—there being four boilers, two in each battery, the batteries being set facing each other so that one fire-room will do for both, the flues could not be carried down at the rear ends of the boilers, even were there room, not considering other objections.

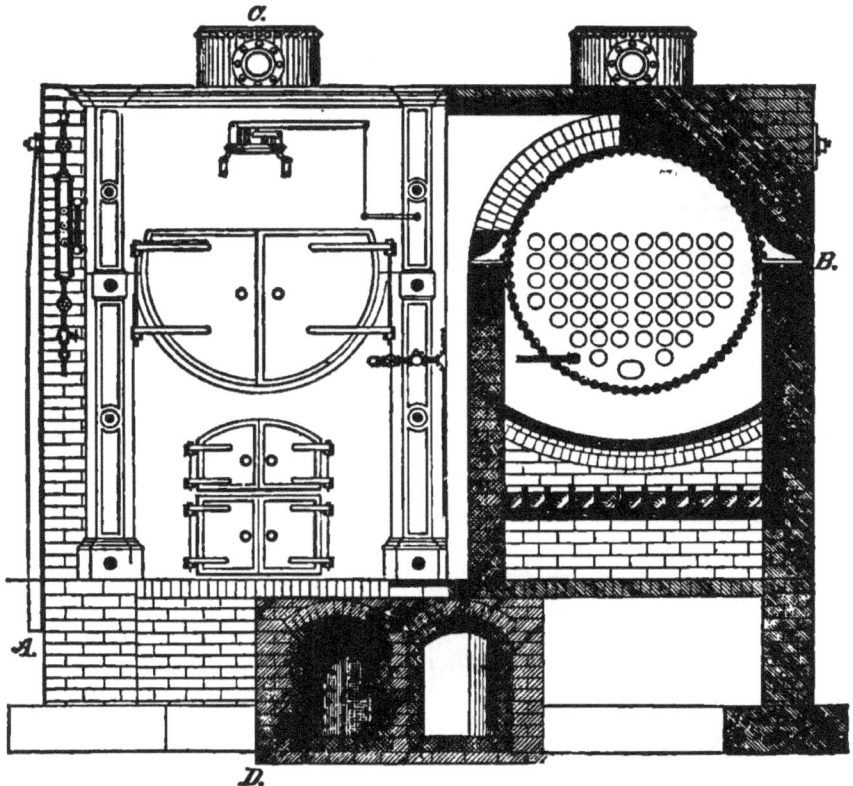

FIGURE 81.—FRONT.

It was decided to make the middle wall of each set of boilers sufficiently thick to have two flues, each 12x36 inches, leaving two courses of fire-brick between the furnaces and the flues. The downward flues from each set of two boilers are carried separately for about ten feet of their length after all the turns are passed, and then connect with chimneys 28x36 inches in the clear and 182 feet high.

It was suggested that the hot draught, after leaving the boilers in its passage through the downward flues, in consideration of the great heat which was expected at this point between the two furnaces, would have a perceptible effect on the draught of the main chimney-shaft, but the preponderance is so greatly in favor of the chimneys that no trouble

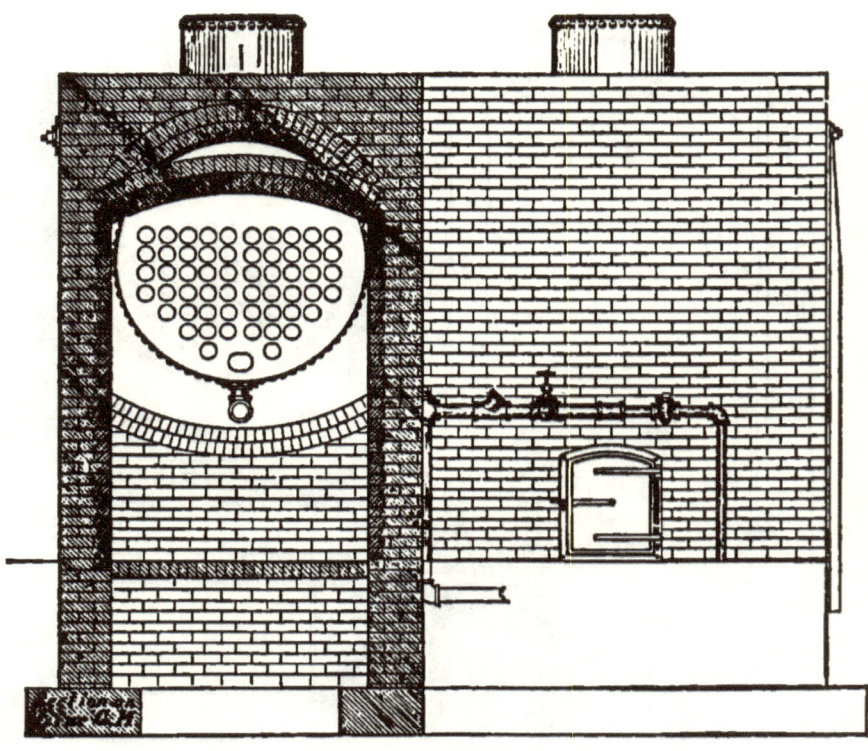

FIGURE 82.—REAR.

has ever been experienced, and the engineer in charge cites, as an evidence of the intensity of the draught, that in moderate winter weather he has been able to keep the whole building (new part and old part) warm with but one of the boilers; which would indicate that a square foot of the boiler-surface makes steam for about fifteen square feet of radiator surface.

MISCELLANEOUS.

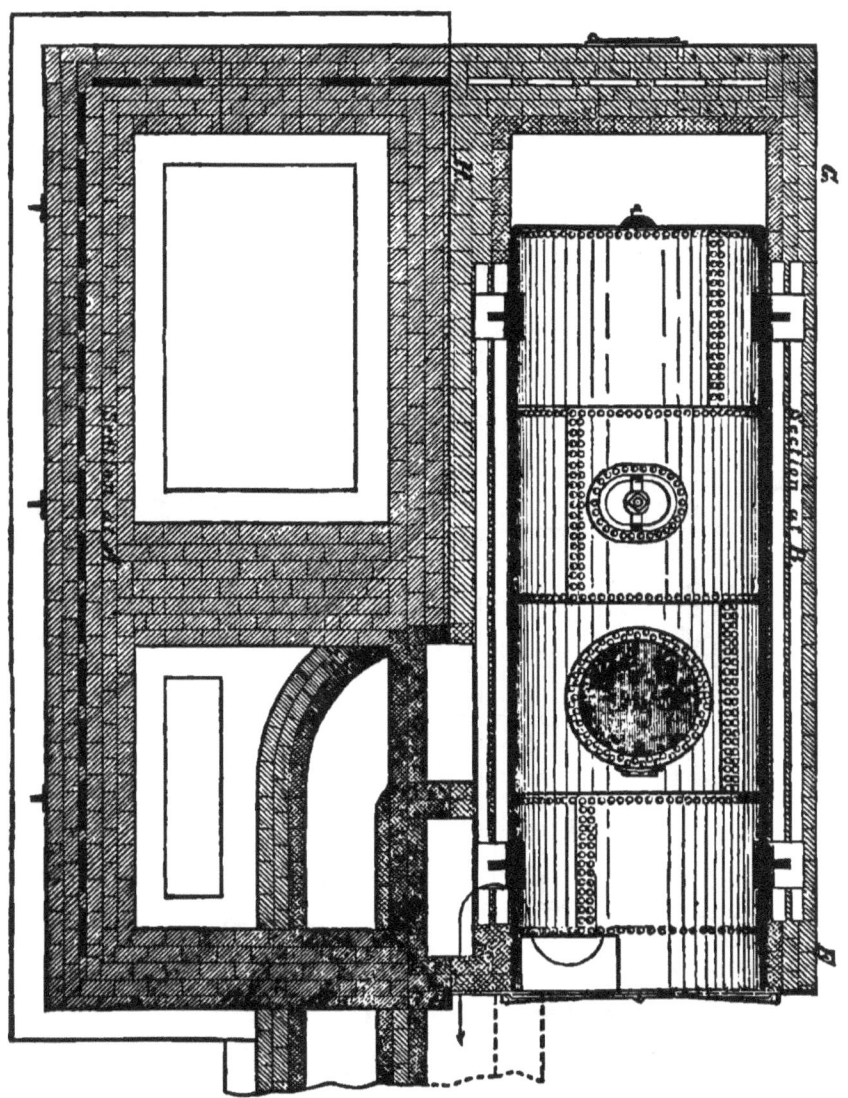

FIGURE 83.—HORIZONTAL SECTIONS.

Where the weight of the centre wall rests on the arches of the underground flues, a very heavy casting corresponding to the plan of the flues is introduced to distribute the weight.

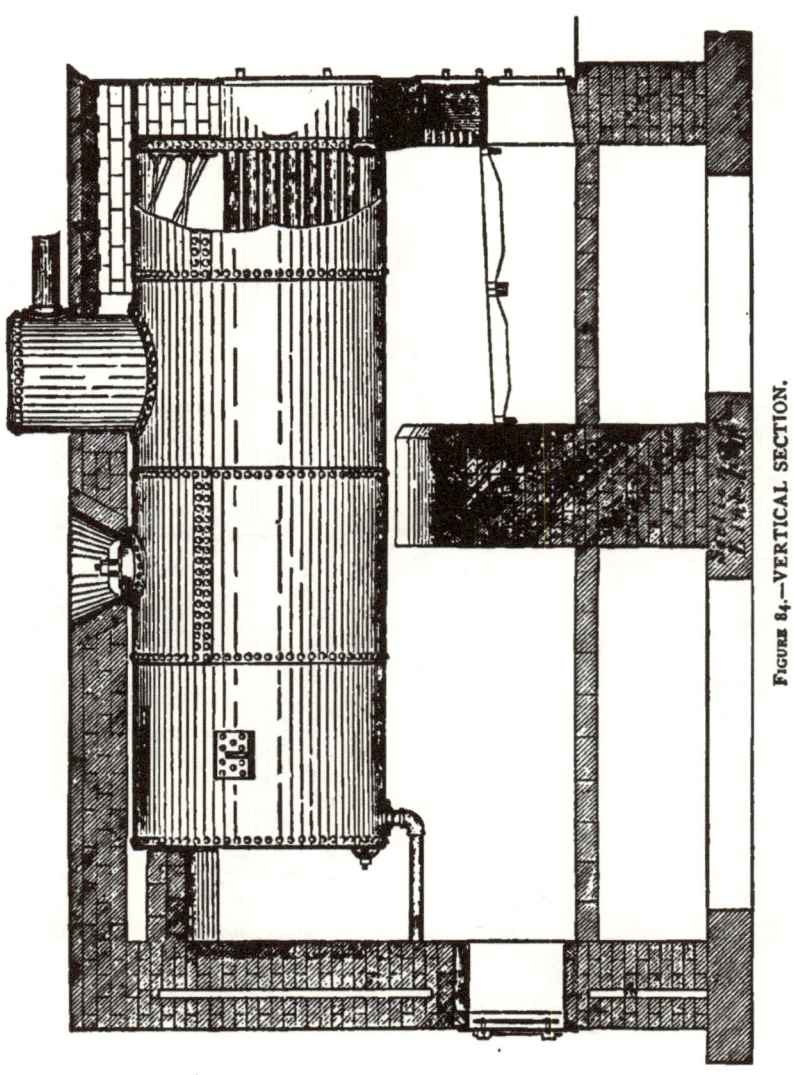

FIGURE 84.—VERTICAL SECTION.

Another novelty which was here introduced by the superintending engineer is heavy bars of iron reaching the whole length of the walls under the boiler-lugs. At the front pair of lugs the boiler is allowed to rest heavily on the irons, but the rear lugs are set on rollers, the object being to prevent the side walls from cracking. The front lugs are the neutral point, and, being close to the front, the expansion forward is only nominal, but the thrust of the boiler backward, which may reach three-eighths of an inch, is thus provided for.

It is sometimes said that one-quarter or three-eighths of an inch cannot make much difference on the length of boiler-walls, but when no provision is made for it, and a crack is once started by the thrust, though the boiler may draw back the brick-work cannot, and the sand and mortar sifts down into the crack and between where the boiler touches, so that when the boiler expands again it will open the crack further, and thus go on.

The boilers are four in number and fifteen feet long between the heads, with an extension of sixteen inches for the front connection. The two new ones are made of steel three-eighths of an inch thick in the shells and one-half an inch in the heads.

WARMING AND VENTILATION OF THE WEST PRESBYTERIAN CHURCH IN NEW YORK CITY.

On the north side of West Forty-second Street, and overlooking Reservoir Park, is the church known as the West Presbyterian. It is seventy-eight feet front by a depth of 140 feet, including the lecture-room. The auditorium proper, outside the chancel, is seventy-two feet square, and the height to the stained-glass skylight (which is thirty-five feet square), in the centre, is sixty-two feet from the floor. The roof-trusses are concealed within four arches, whose centres are fifty-four feet high, and which intersect each other in such a manner as to give the auditorium and galleries the effect of being under a dome. The number of persons that can be seated comfortably is 1,200. It has recently been refitted and altered, and provided with the system of heating and ventilation shown in our illustrations, under the plans and direction of Messrs. J. C. Cady & Co., architects, of New York.

Originally the building was without wall-flues. To remedy this defect advantage was taken of the pilasters, or continuations downward of the truss-arches (see plan). They were widened on their faces below

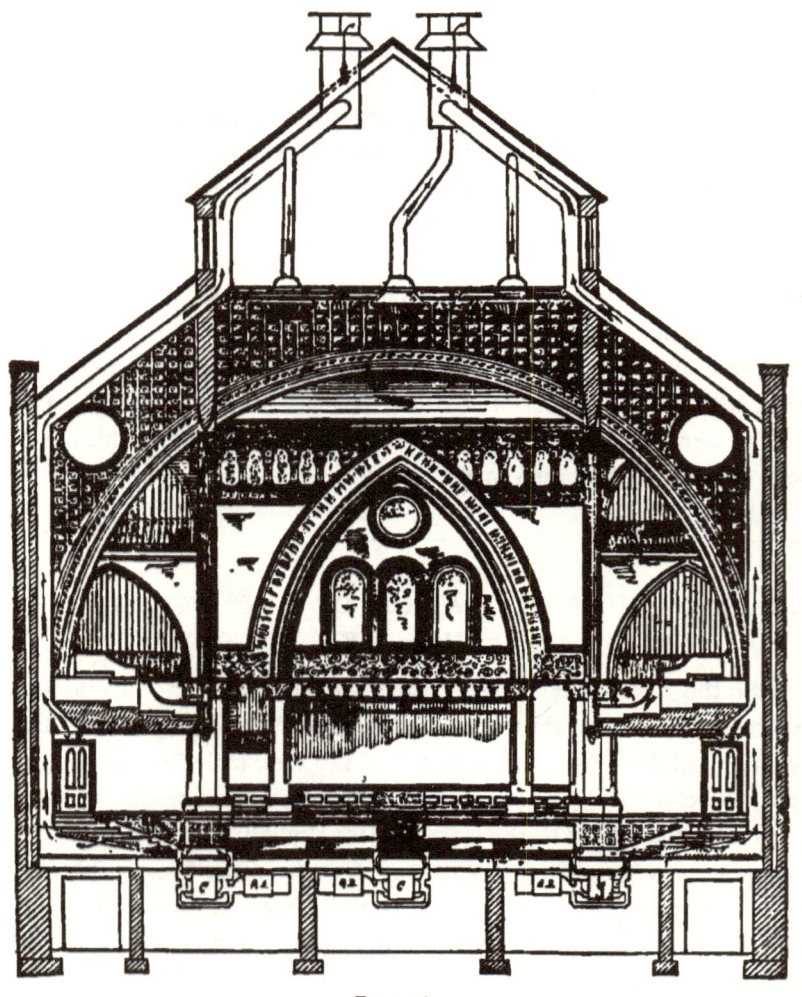

FIGURE 85.

the galleries until a flue 12x16 inches was formed on each side of them. Above the galleries these flues are carried against the walls, and in the corners formed by the arches and the roof, and terminate in galvanized-

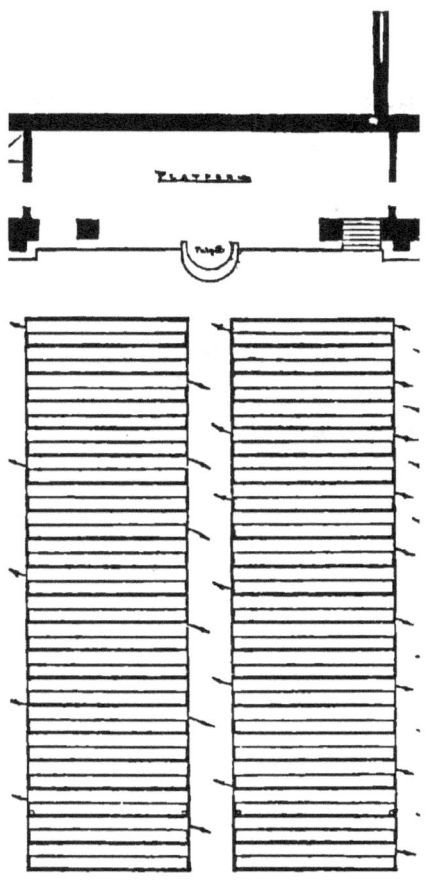

iron flues in the raised part of the roof, where they are collected into two large ventilating-tops, one for each side of the building.

The number of flues thus formed is twelve, each 12″x16″ in cross-section, making sixteen square feet of outlets, not taking into consideration five 12-inch round flues, one from each of the five chandeliers under the skylight.

In each flue is a Bunsen gas-burner, surrounded with a sheet-iron tube, to produce rarefaction of the air at times when steam is not used in the building. These flues are marked V R on the auditorium plan (Figure 86), and are seen in section in the interior view, Figure 85. The principal opening into each is through 20″ x 24″ registers in the gallery ceiling, but the spaces underneath floors of the pews on the sides are also connected with them, registers being placed in the risers of the steps in the cross-aisles.

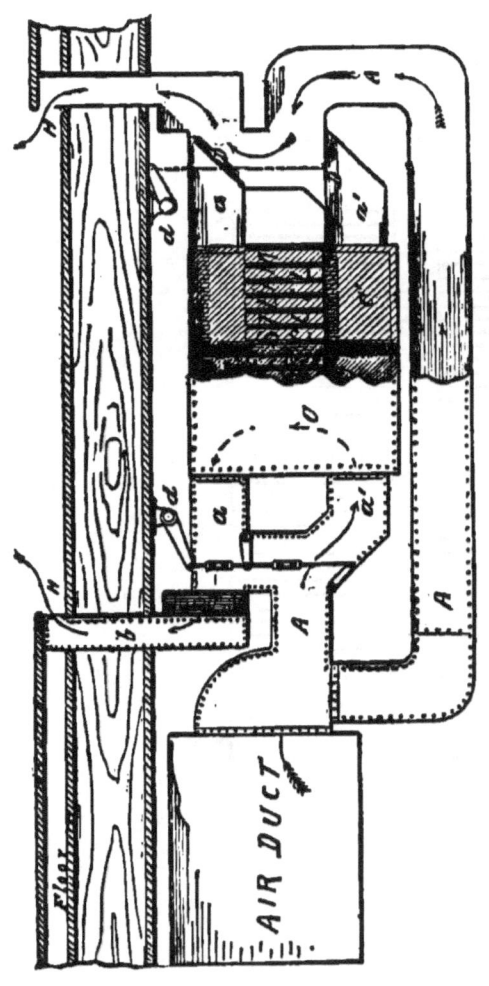

FIGURE 87.

As a further means of warming these flues, and also for getting rid of the products of combustion from the clusters of gas-jets, each cluster is hooded, as shown in Figure 88, and the pipe from the top of the hood carried into the nearest flue.

Fresh air for the building is taken from the roof through a shaft 6'x6', which has been constructed through the room back of the pastor's study. This is a brick shaft which runs to the basement, and from which the horizontal galvanized-iron air-ducts take their supply. Three lines of indirect radiators, with their air-ducts, are used in the basement under the three principal aisles. To avoid registers in the floors and to secure a finer and more subdivided admission of air and heat, the floor underneath the pews (but not in the aisles) was raised about five inches.

Through wire gratings, H, in the risers of the steps thus formed, the air is admitted to the church, as shown by the arrows, Figure 86.

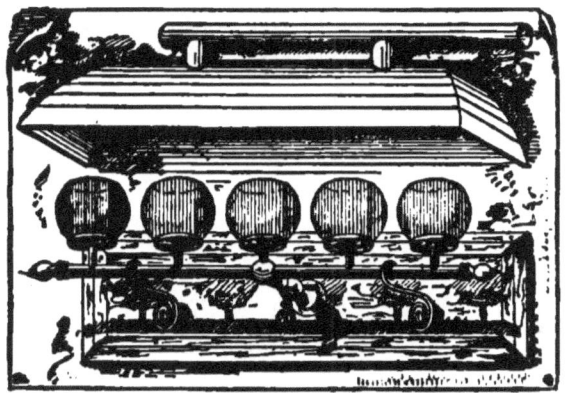

FIGURE 88.

Below are clusters of steam-coils, four to each aisle, which are divided in the middle so as to give a separate supply to each side of the same aisle. These radiators are inclosed in galvanized-iron cases and arranged so that hot air or cold air, or a mixture of the two, can be passed into the church at pleasure.

Figure 87 is an enlarged detail of the heaters, C, in Figure 85. Fresh air from the air-duct passes through the pipe A to the coil-casing. When it is desirable to pass all the air through the radiator the dampers or valves D are in the position shown to the left of the cut, but when cold air only is to be admitted to the church the dampers are arranged as shown at the right. To secure different degrees of heat the dampers in the pipes A and A' are both partly opened, allowing part of the air to go through the coil and part directly into the pipe B.

At two places within the auditorium, and near the side doors leading from the vestibule, all the "switch-valves" or dampers in the air-pipes can be operated. The rods $d\ d$, Figure 87, connect with the damper-levers, as shown, and in turn connect, by bevel gearing, with a shaft which extends across the front of the basement. This shaft, in turn, connects with an upright which extends through the church floor and is topped with a hand-wheel. By the manipulation of this wheel the attendant can admit more or less cold air and keep the church at a constant temperature.

The steam-heating and ventilating contractors were Messrs. Baker, Smith & Co., New York, and the architects Messrs. J. C. Cady & Co., of Trinity Buildings, New York.

PRINCIPLE OF HEATING-APPARATUS, FINE ART EXHIBITION BUILDING, COPENHAGEN.

The following are two systems of warming proposed by A. B. Reck, of Copenhagen.

Figure 89 represents the principles of an apparatus stated to be in use at the Royal Fine Art Exhibition Building, Copenhagen. It is so arranged as to work with live steam from the boiler or the exhaust from electric-light engines, or a mixture of both. In the figure, F is a pump to return the water of condensation to the boiler. D represents the engine, some of the power of which operates the pump. Steam leaves the boiler through valve V, passing through the engine, thence through the valve a' to the heating pipes, which of course have a pressure but little above atmosphere in them. Should the quantity of steam passed by the engine be greater than the requirements of the heating apparatus, the back-pressure valve T opens, allowing the excess to escape to atmosphere. But, on the other hand, should the engine be performing small duty, and passing little exhaust steam, the regulating-valve R on the boiler opens automatically and passes the extra quantity of steam necessary to maintain the desired fullness in the heating pipes through the stop-valve a^2.

Figure 90 shows an apparatus the principal claim for which is the exhaust steam from engines run at night for lighting purposes can be used to heat water and be made to retain the heat in a non-conducting coil-case for use next day without running the engine, and also that the water can be circulated by a pump throughout the system, while exhaust steam cannot be carried to such distances. In many

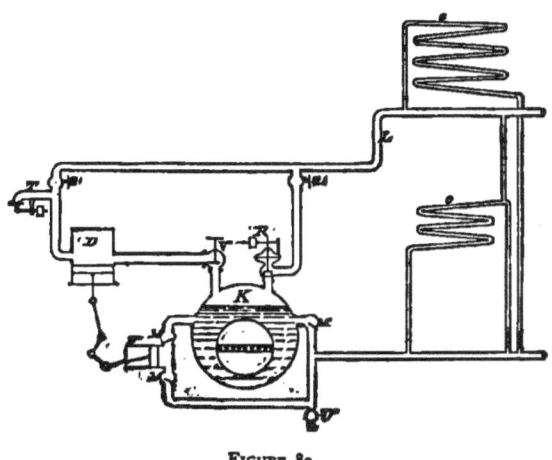

FIGURE 89.

respects our remarks on Figure 89 will apply here, the same letters applying to similar parts. It will be observed that the pipe I conveys the steam—live or exhaust—to the coil S within a water-tank (P), the condensation being returned by the pump F. The other pump, C, circulates the water from the tank P to the smaller reservoirs O throughout the house, storing hot water in the same and bringing all the water to as high a temperature as possible during the night, or at such times as the engine is running. When the cases which inclose the heat-reservoirs O O are closed at their tops, no passage of air occurs, hence there is only a comparatively small loss of heat; but where heat is required the registers are opened, and the warm air

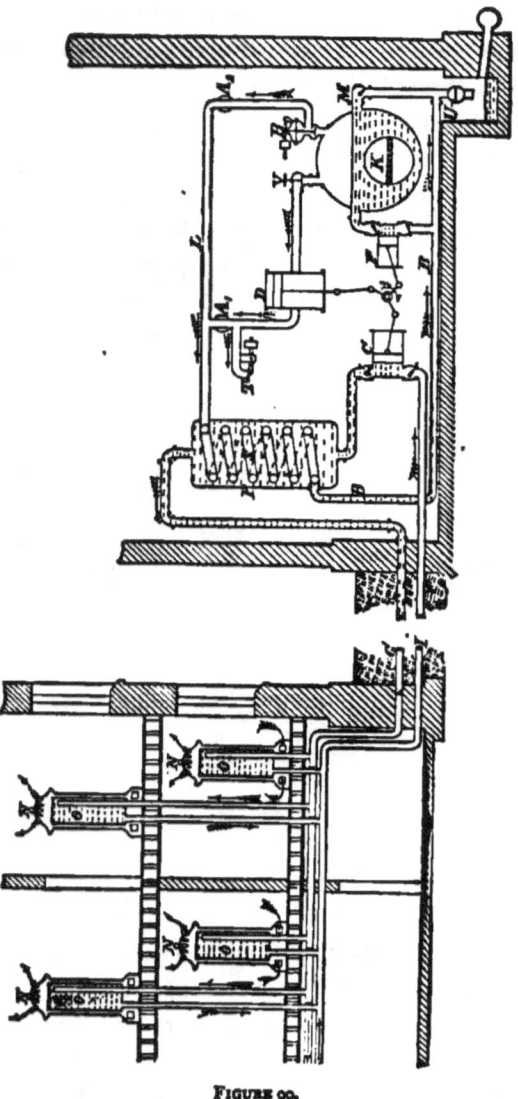

FIGURE 90.

circulates freely. Other points will suggest themselves to the reader by a study of the diagram.

WARMING AND VENTILATING THE OPERA HOUSE AT OGDENSBURG.

THE building, though known by the name of "Town Hall," is, in reality, the municipal building, jail, town hall, and opera house combined.

The auditorium of the theatre is 60 feet wide by 68 feet long and 50 feet high, with two galleries, and will seat 976 persons.

The stage at the curtain is 35 feet wide and 46 feet deep, with a height of 60 feet.

The system of warming is by direct and indirect radiation combined, and the ventilation is produced by a *vacuum movement;*

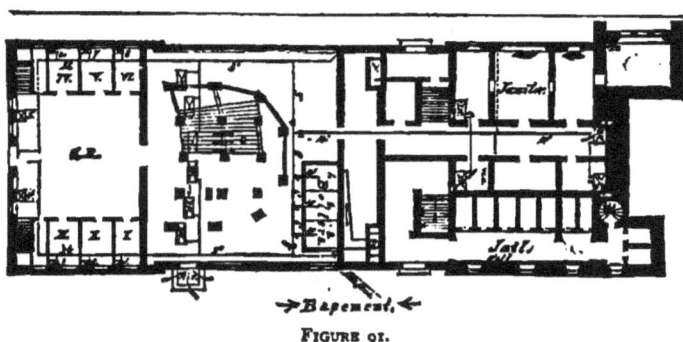

FIGURE 91.

aspirating-shafts exhausting the air at the levels of the floors and ceilings of the main floor and galleries.

In cold weather the aspirating-shafts are assisted by the natural condition produced by the dense cold air forcing its way to the inlet-ducts.

The warmed air in winter and the fresh air in summer is admitted at 144 places through the auditorium floor (see arrows, Figure 92). These openings are nine inches long by three inches wide, with register-faces set into them, having an aggregate area of twenty square feet of openings.

The fresh air is taken in at the side of the basement (marked *air* on basement plan) and drawn down through a short shaft; from thence it is delivered to the large boxed coils.

The warm air from these coils is then passed between every third joist (which radiate and slope out and upward from the stage) and is delivered through the risers of the steps upon which the seats are fastened.

The space between the joists through which the warmed air is drawn is lined with a heavy tin casing, flanged through the risers,

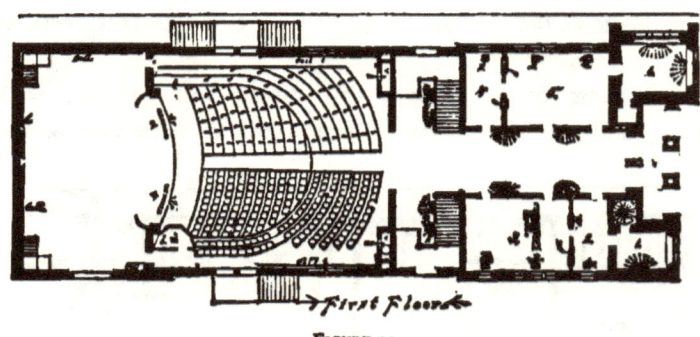

FIGURE 92.

before the register-faces are inserted. A detail of how this is done can be seen in the perspective drawing (Figure 94).

The direct radiation is by long coils placed along the outside walls and under the windows of the main floor and galleries, to prevent downward cold currents at the walls.

The aspirating-shafts have an area in cross-section of thirty-seven square feet. The shaft c is warmed by a box-coil, placed just above the level of the first gallery, the coil being made to fit the space, so that the air in its upward movement would all have to come in contact with the hot pipes. The shaft on the other side of the building is warmed by the smoke-pipe from the boilers passing through its whole length.

The obstruction to the shafts caused by the heating furnaces reduces their capacity to about thirty square feet. Six registers, 21x29 inches, open into each of these shafts, giving an area, when allowance is made for the fret-work, of about thirty-five square feet.

We are told the velocity of the air at the registers will average 420 feet per minute, which must be equivalent to the moving of the air once every fourteen minutes, or about fourteen cubic feet of air per person per minute.

In addition to the aspirating-shafts, there is in the centre of the ceiling, over the chandelier, an arrangement of nicely-disguised openings, aggregating fifty square feet. The air in its escape through these openings is controlled above the ceiling by trap-doors operated from the wings.

The stage is warmed entirely by direct radiation, the surface being very large and divided into sections, with a view to keeping the stage and the auditorium at as near the same temperature as possible.

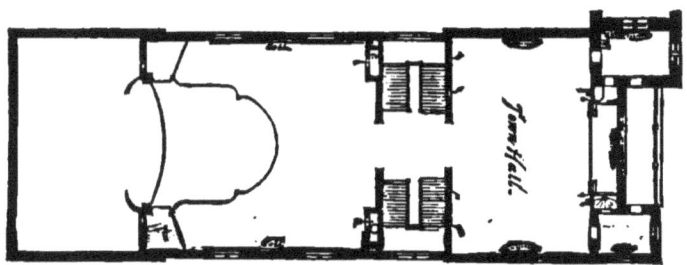

FIGURE 93.—SECOND FLOOR.

Each private box has its own warm-air register, and there are two large direct radiators under the front of the stage, the air circulating from the auditorium in and out through fret-work.

The Town Hall, 64x48x34 feet, on the second floor, is warmed by three direct radiators and by four indirect radiators, with the air delivered in six places.

The four principal offices on the first floor are warmed by indirect radiation, and the rest of the building, including halls and jail, by direct radiation.

Steam is furnished by three of the "Nason" sectional steam-generators, two of them being always sufficient for the building, with one in reserve.

198   STEAM-HEATING AND STEAM-FITTING PROBLEMS.

The apparatus is what is known as a "gravity return," and will work at all pressures.

The size of the mains is shown in the basement plan (Figure 91), with the warming of the green-room and dressing-rooms, which is done

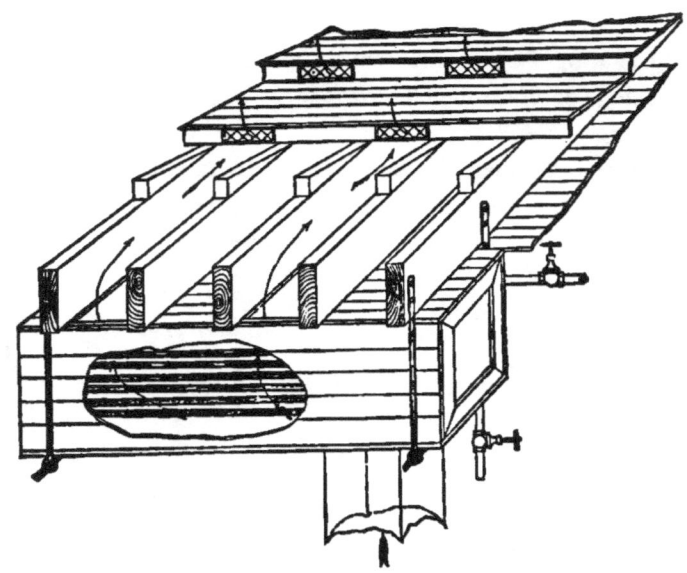

FIGURE 94.—DETAIL ILLUSTRATING METHOD OF ADMITTING WARM AIR TO AUDITORIUM.

by overhead pipes. Other details may be gleaned from a study of the diagrams.

The architect of the building was G. A. Schellenger, of New York

SYSTEMS OF HEATING HOUSES IN GERMANY AND AUSTRIA.

IN the greater part of Germany and Austria the system of heating the houses is, to a great extent, insufficient, uneconomical, and inconvenient, being, to a certain extent, a mere modified and improved form of the method of the ancient tribes of heating large stones or other large bodies in their fires and allowing them to cool off in the rooms or huts, thus heating the air. The stoves almost universally employed in most private, and even public, houses are of porcelain, more or less ornamented, of rectangular cross-section, and quite high, containing in them a zig-zag flue, the fire-place being at the bottom. The fire heats the long flue in the stove itself, which then heats the air of the room.

The disadvantages are numerous. The walls of the stoves, being made of one of the best non-conducting materials, require a great amount of heat, and for quite a long time, before they conduct the heat through them to the air of the rooms, so that most of the heat of the walls passes through the flues into the chimney, and, of course, is lost. Owing to the zig-zag form of the flues, the draught at starting is very poor, and the stoves, being made of many separate pieces, frequently contain many cracks, so that the room is often filled with smoke and disagreeable gases ; but as this is a daily occurrence one is expected to become accustomed to it. It is still more disagreeable, as the coal used resembles our bituminous coal and gives off much more disagreeable gases if not completely burned.

FIGURE 95.

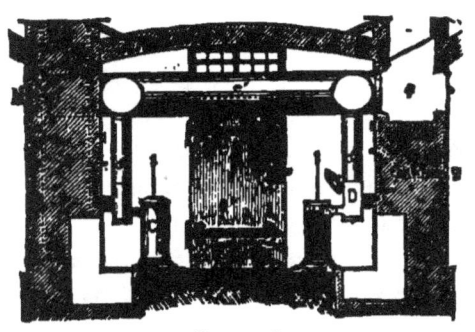

FIGURE 96.

A further disadvantage is that the fire has to be lighted every morning, and that it requires several hours, sometimes, before the non-conducting material of the stove is sufficiently heated to warm the room. For a man in business this is especially inconvenient, as he either has to be aroused very early in the morning by the servant starting the fire and by the smoke, or else the room is cold for the few hours in the morning while he is using it, and when he is about to leave it, it will be beginning to become warm—or, in other words, it is cold while he uses it and warm in his absence.

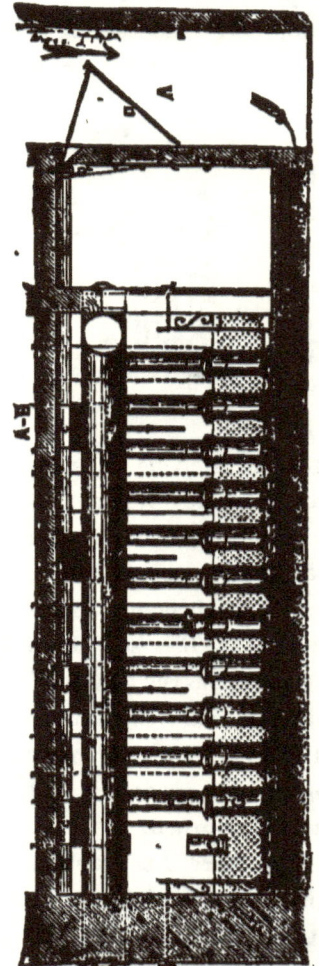

FIGURE 97.

Every room has to have its own stove. Hallways and bath-rooms (when there are such) are seldom heated, and therefore have the temperature of out-doors, so that one frequently takes cold in going from one room to another. All doors must be kept closed, thus preventing all ventilation by means of hallways, stairs, etc. The fire being allowed to go out as soon as the walls of the stove are warm, there is no ventilation by means of the draught of the fire.

The heat in the walls of the stove after one firing is said to be sufficient for the day. The heat of the inside of the flues, which is much greater than that of the outside, is lost in the chimney after the fire is out, as there are always enough cracks or openings to allow a slight draught.

Another objection is that the stoves take up much room and require a very solid foundation.

American stoves are sometimes used, but apparently are not liked, as the "regulirung" (draught-regulation) seems to be far beyond the conception of the average German servant.

Another method, known as "Hauber's Patent," has been introduced in many places, especially in Munich, in the last few years, and seems to be, both theoretically and practically, one of the cheapest and most economical methods.

Before explaining it, it may be well to call attention to the fact that in most furnaces or stoves the fresh coal is thrown *over* the hot walls, and is therefore heated in a flame containing little or no free oxygen. The consequence is that large quantities of combustible gases, rich in heating qualities, and a large amount of black smoke containing unburnt carbon, escape unused into the chimney, carrying with them heat, thus representing a great waste. If these gases could be burned while hot there would be considerable saving, besides avoiding the disagreeable black smoke. This could be done if the fresh coals could be placed *under* the hot ones, so that the gases given off by them are burnt while passing through the hot coals and while the gases contain free oxygen. This is the principle of the system to be described and seems to be accomplished.

A single stove, or "element," as it is called, is shown in about one-twentieth size in the accompanying cut (Figure 95). It consists of a cast-iron cylinder C, open at the top and containing at the bottom the contrivance d'', which is like an inverted frustum of a cone, perforated so as to allow the air to enter, but preventing the coal from falling out. D is the flue fastened to the wall and ceiling, forming a great portion of the heating surface; $b\ b$ is the fixed cast-iron base. The cylinder C fits loosely on the base $b\ b$ and at e, so that it can be taken away and replaced by another without requiring any fitting other than that the opening e should be opposite that in the flue D. The "elements" or cylinders are filled to the top with coal in the coal-bin, and carried by means of a handle at the top into the rooms, thus serving at the same time as coal-bucket and stove. A small charge of wood and paper or shavings is placed on the top of the coal. When the room is to be heated it is only necessary to light the paper and put the lid on somewhat slanted, so as to allow the draught to enter at the top until the coal is ignited, after which the lid is closed and the draught at the bottom regulated according to the amount of heat required.

As is easily seen, the gases from the fresh coals will have to pass over the hot ones, thus burning them completely, and thereby avoiding smoke and utilizing all the heat of the coal. No attention is necessary until all the coal is consumed, when the cylinder is replaced by a filled one.

The joint at *e* need not fit tightly, as the section at that point is small as compared with that at the flue, so that there is a greater velocity of the gases at that point. Even if the stove is placed as much as half an inch from the opening no gases will come out at that point if the draught is good.

FIGURE 98.

The advantages are manifold. It is very economical, requiring, as it is claimed, only twenty-five to sixty per cent. of the amount of coal used in the ordinary systems. There is no smoke or incomplete combustion, and therefore no such loss of heating material. There is very little attention required, and no dust from coal or ashes in the rooms. It is very easily regulated by a valve at the base. The stoves, being simple in construction, are comparatively cheap.

This system is used principally for large buildings, shops, schools, hospitals, etc., where it seems to be very satisfactory. It is also frequently used for drying-rooms where very high temperature is required.

For heating a large building, a so-called "battery" of stoves is placed in an apartment in the cellar, from which the hot air passes by means of pipes to the rooms. Such a chamber is shown in sections and plan, in Figures 96, 97, and 98.

The following data were given to the writer by the Austrian Company: A charge of coal weighs about 10 kilog. (22 lbs) and burns from six to fifteen hours, according to the amount of heat required. On the average two fillings a day are required. This, at the rate of $4 per ton, would be about four cents per charge. For heating and ventilating, in which the air of rooms is renewed three times per hour, one element with bituminous coal will be required for every 85 to 90 cubic metres of space heated, and for brown coal, 60 cubic metres; for hallways, one element for 100 to 120 cubic metres.

The cost of an element in Austria is about $12 (30 fl.) ; for heating by means of chambers in the cellars, about $20 to $24 per element, including all fixtures, but not the erection.

THE STEAM-PIPES IN NEW YORK STREETS.

A CORRESPONDENT writes: "As you must be aware, there is a great difference in matters of detail between the two steam companies of New York City, an approximate sketch of each of which systems I submit—Figure A being the American and Figure B the New York Steam Co. In general, both send out steam through wrought-iron mains, and the New York Co. is receiving back a large quantity of its condensed water (all that is not used in the engines) through a parallel main, and the American company has parallel mains through which it intends to receive back what water it can when the company is in working order.

"System A has three pipes, two (1 and 2) being the conveying pipes, and one (3) the return pipe, but in many places the conveying pipes are supplemented by distributing pipes, which start from the flanges (f) at the junction and run parallel with the other pipes, while in other places the conveying pipe does duty for both purposes. Pipes 1 and 2 are connected with the 8-branch cross (junction) with a stop-valve c, and an expansion-joint d, the expansion of the metal of the pipe which

204 STEAM-HEATING AND STEAM-FITTING PROBLEMS.

takes place between any two street-corners being forced within the *slip-joints*. On top of the steam-junction J is placed a smaller juncture (J)

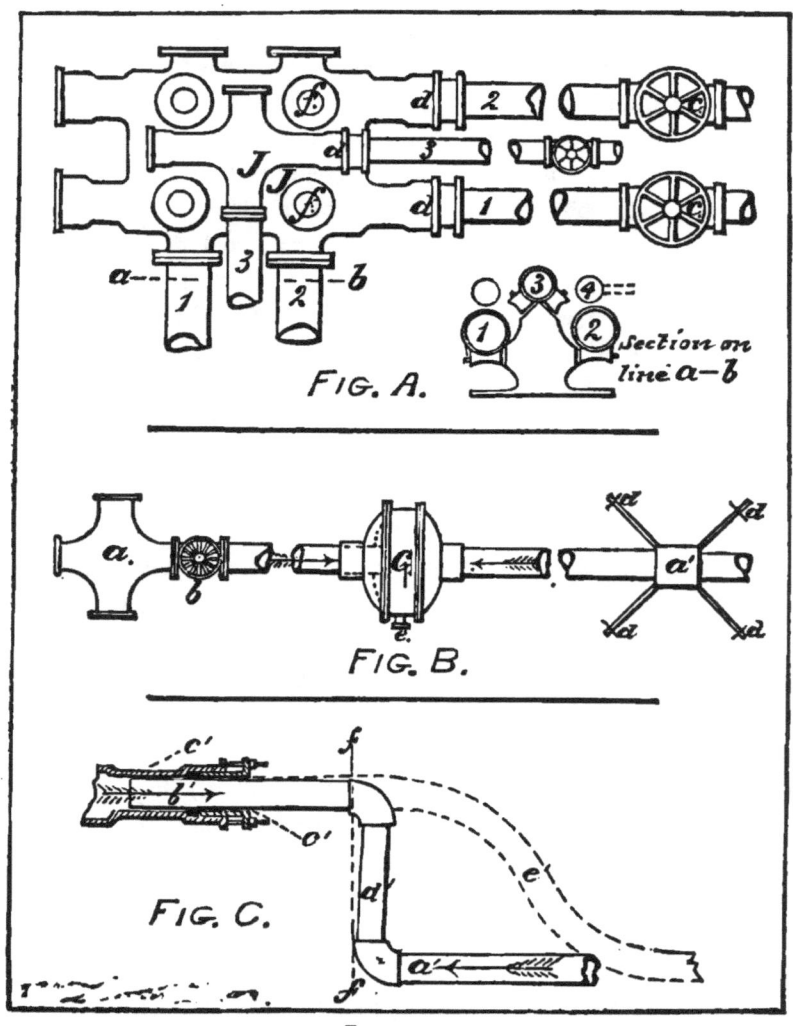

FIGURE 99.

for the return water-pipes, to and from which run the return-pipes (3), with stop-valves and expansion-joints also. Now, in the majority of cases, if not in all, the return-pipes where they leave the "return" junction, just beyond the expansion-joints, drop a certain distance, to get down to the position shown in the elevation—elbows and nipples, or short pieces being used to make the necessary offsets. It is here the principal trouble seems to be, with the offsets and expansion-joints, though it may be with the expansion-joints of the steam-pipes as well. When expansion takes place, the thrust of the long pipe, caused by expansion when heated, is against the lower elbow, and the resistance from the pressure within the pipe is against the upper elbow, the thrust not being in a right line, but angular, causing the brass of the slip-joint to bind on the sides of the sleeve. The sketch C will give an idea of this in a better manner than words can. The thrust of the pipe is in the direction of the arrow a', and the resistance from the pressure and from the friction of the gland is in the direction of the arrow b', which causes the offset to assume the position shown, binding the brass at c and c' in such a manner that it is almost impossible to move it, the result being the fracturing of pipes or elbows. The dotted lines e' shows the bent pipes (offsets) which were substituted in places for the elbows and piece d'.

"With the New York Steam Company the result is different. They do not attempt to force the expansion from street-corner to street-corner (the difference in length for 200 yards being about 15 inches). They put in an expansion-joint every 75 feet or so and anchor the pipe in the middle of the distance between the joints—a conception of which you may be able to get from the diagram B : a, is the junction ; b, the stop-valves ; c, the expansion-joint ; $d\,d\,d\,d$, the anchorage. If the distance from the anchorage to the expansion-joint is 37½ feet, the greatest movement on the diaphragm of the compensator cannot exceed one inch for a range of temperature from zero to the temperature of 100 pounds of steam. The return-pipe is run in the same manner as the steam-pipe.

"The duplicate system used by the American company presents features not covered by the New York company's method, and might be used as an argument of their inventions *not to do things cheaply*, but it is questionable if what can be gained in the way of preventing interruptions will warrant the expense and complication where the street blocks are as short as in the lower districts."

Later, a committee appointed to look into the cause of the frequent bursting of the American Company's pipes decided that the principal reason for giving way was the water-hammer, caused by steam meeting water held in depressions of the pipe. Pipes that were ruptured in their lengths, having bursts similar in appearance to frost bursts, were tested at their remaining sound parts and found to stand a pressure of about 1,100 pounds per square inch, from which some estimate may be formed of the violence of the blow that may be given in a four-inch pipe—under the proper conditions—by a pressure of about 60 pounds per square inch.

SOME DETAILS OF STEAM AND VENTILATING APPARATUS AS USED ON THE CONTINENT.

We are indebted to "Hygiene des Habitations," a pamphlet published by MM. Geneste, Herscher et Cie, of Paris, for the accompanying details of apparatus used by them in the warming and ventilating of buildings in France and other continental European countries. They

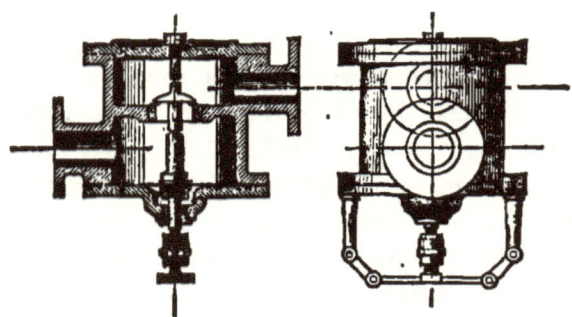

FIGURES 100 AND 101.

use water-tube boilers, as they consider them less liable to disastrous explosion, and claim an advantage for them on account of the shorter time it takes to get up steam over boilers which contain a large quantity of water. From the generator they carry the steam to the highest part of the premises, and supply it downward for distribution to the coils.

MISCELLANEOUS. 207

Sufficient pressure is carried in the boilers to work pumps for the return of the water of condensation from the receiving tanks and to work engines to drive the fans.

Before carrying the steam to the top of the house it is reduced to a very low pressure by the use of a "regulating-valve," which is shown in section and elevation at Figures 100 and 101. To any one acquainted with the regulating or reducing valves used in this country the cuts speak for themselves and require no explanation from us

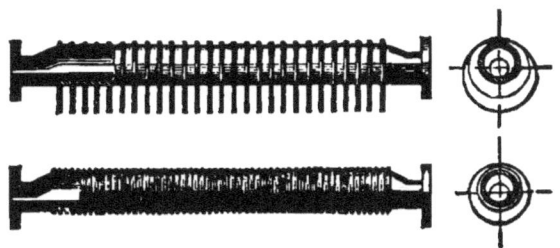

FIGURES 102 AND 103.

From the point where steam is first carried to the top of the house its movement and the movement of the water is in the same direction— that is to say, in the direction of gravity.

They use air-purging cocks on the heaters—automatic air-vents— on the differential expansion principle, in which a volute coil of two metals open and close the outlet. For steam-traps—*purgeur à contrapoids*—they use a counterbalanced metal float, which is solid, or nearly so, and which is but partly balanced by a counterpoise that is always above the water-level, but which leaves preponderance enough in favor of the part that will be submerged when the water of condensation rises to have it act as a float, the object presumably being to get a float that cannot be collapsed or filled with water under pressure.

The class of *extended* surface-heaters used is shown in Figures 102 and 103 in part section and elevation. For heating ordinary rooms they recommend direct radiation at the cold points of the room, and consider Figure 104 a good arrangement, and say: "Place the radiating surfaces with the extensions downward in the allayings of the windows, the fresh air being let in by an aperture contrived in the allaying itself." The upper part of the radiator is for direct heating only, and the lower part is for warming the incoming air.

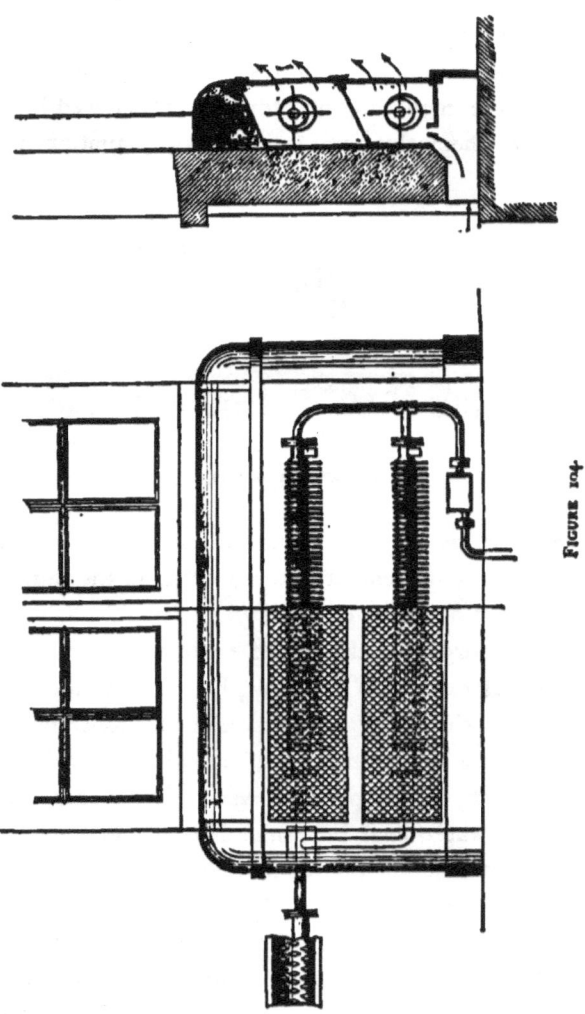

Figures 105 and 106 show the arrangement of a coil for hot-water heating and the means of admitting air behind it, but not in contact with it, the air of the room which circulates through the heater and is warmed thereby mixing with the cooler (entering) air as shown.

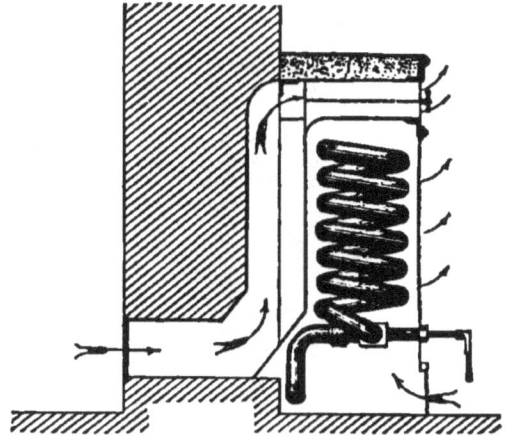

FIGURE 105.

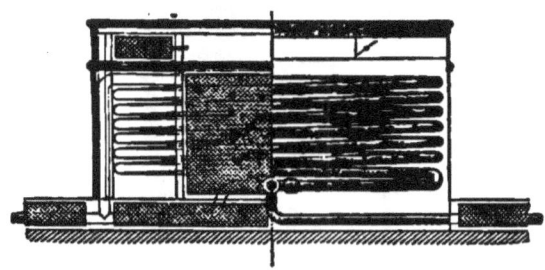

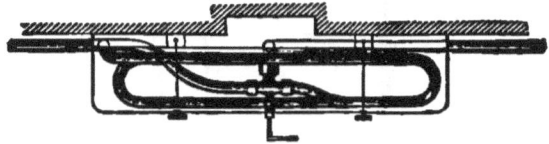

FIGURE 106.

It will be noticed by a study of the figures that the hot-water coils are in a continuous circuit, and that by the turning of the three-way cock the water is made to circulate through the coils or past them, as desired.

In forcing or exhausting air by mechanical means two differ-

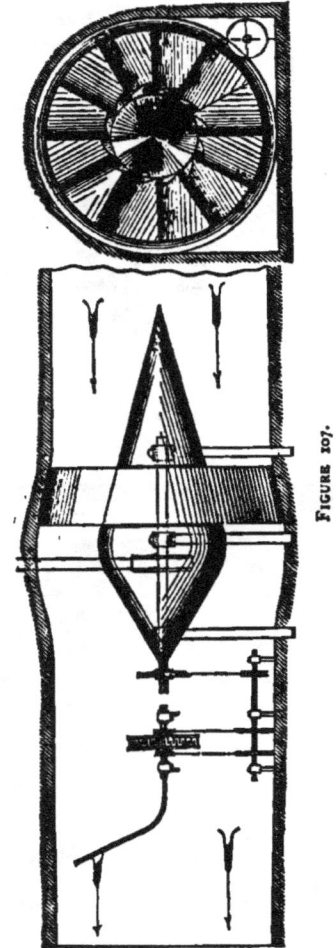

Figure 107.

ent classes of fans are used. The one is the ordinary centrifugal fan (L. Ser system), and the other "helicoidal" (Figure 107), or what is sometimes called the "propeller" principle in this country. This fan is used when large quantities of air with very high pressures is required, and is considered ample for usual premises.

Figure 108 shows a small fan of this kind, called a *hydro-ventilator*, it being driven by a small water-motor, and usually used as a supple-

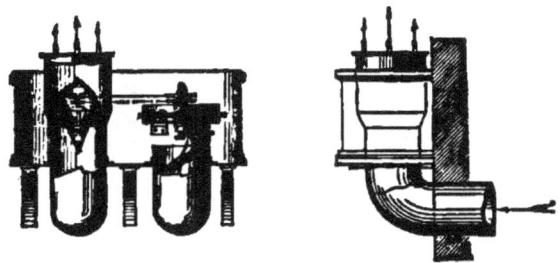

FIGURE 108.

mentary fan at some part of a building where the local currents want assistance, the trouble and annoyance due to belts and most other means of transmitting power being largely overcome by the water-motor, which requires only a supply and a waste pipe.

MISCELLANEOUS QUESTIONS.

APPLYING TRAPS TO GRAVITY STEAM-APPARATUS.

Q. HEREWITH I send you a sketch (Figure 109) of the two boilers in our heating-apparatus, showing the manner the steam-pipe A leaves both boilers and the point at which the return-pipes *b b* and the main return-pipe B enter the boilers again. The apparatus does not return the water of condensation properly at all times, and I am considering

FIGURE 109.

the question of applying a return-trap, but I wish to do it in such a manner that I may return either by gravity or by the trap. What way would you suggest as the best and simplest, or what is the usual method of doing it?

A. We do not know that the trade has any fixed method for applying a trap to a gravity apparatus.

In your case we would cut the return-pipe (B) at C, and introduce a stop-valve with a *tee* on each side of it, as shown by our alteration in your drawing. To these *tees* we would attach stop-valves E and D, and to the valve D a receiver. From this receiver then run the usual pipe F to the trap, as shown; then from the trap take the usual discharge-pipe G, and instead of making extra holes in the boiler-heads, return it through the valve and *tee* E to the main return-pipe again. To operate the apparatus by the trap alone, close the valve C, open the two valves E and D, and operate the trap in the usual manner.

It may be well for us to say—though it may be already known to our inquirer—that the pipe H which supplies live steam to the trap must be taken from the domes of the boilers and not from the main steam-pipe.

EXPANSION OF BRASS PIPE AND IRON PIPE.

Q. Will you kindly inform a lead-pipe plumber what the expansion is of brass pipe through which hot water is passed—also, is it greater or less than for iron pipe?

A. The expansion of brass is greater than of iron pipe for the same number of degrees warmed. It will vary slightly with different compositions of brass, but $\frac{1}{100000}$ of the length of the pipe may be taken as the mean of its expansion or contraction of each degree Fahrenheit it is warmed or cooled between 32° and 212° Fah., while wrought-iron expands or contracts $\frac{1}{150000}$ part of its length for the same conditions.

Example for brass pipe 100 feet long warmed from 40° to 180° Fah.: $100' \times .00001 \times 140° = 0.14'$ (or 1.68 inches) for the amount which the 100 feet of brass pipe expands.

CONNECTING STEAM AND RETURN RISERS AT THEIR TOPS.

Q. In looking over the "Thermus" articles I find No. IX. is devoted to the question of connecting steam and return risers by circulation-pipes at their tops. I would like very much to know what advantages there are in the use of such pipes in a single return system. I can readily understand why they should not be applied to the separate return system, but can see no special use for them in any case.

A. Imagine a rising line, composed of a single steam-pipe and a single return-pipe, unconnected at the top, which will supply steam to and bring water from the heaters of a vertical line in a building three or four stories in height, the heating apparatus being a closed one—that is, a gravity-return or a direct trap-return. If, now, all the heaters on this line happen to be closed at the same time—a not unusual thing in the early fall or late spring weather—the pressure from the main return-pipe will allow the water to "back up" within the vertical return-pipe to a height of about $2\frac{1}{4}$ feet for each pound of pressure per square inch the boiler may have on it. If the boiler has forty pounds of steam up, this is equivalent to sending up the column in the return-pipe ninety feet to the five or six stories of an ordinary building. If, now, some one on the lower story or any story below the fifth, turns on their steam, they establish a steam circuit between the steam and return risers, and all the water above the junction of the radiator so turned on within the vertical return-pipe has to fall or come down past the end of the hot circuit thus established. This water will condense the steam that thus passes in its fall and cause water-hammer and noise.

Another point in favor of connecting top of rising lines is, it prevents water from "backing up" into the lines when the heaters are shut off that is required in the boiler, and in pipe or sectional boilers, or any boiler of small water capacity, this is a serious matter. In many buildings it may not be necessary to so connect lines, but in all apartment-houses or dwelling-houses with closed circuits it will be found to be of advantage. When the lines are so connected the pressure of the steam passes into the return and holds the water down at all times.

The idea that this will prevent circulation within heaters is erroneous, for at the most it can have no other effect than that produced by having steam on the top radiator of a line.—THERMUS.

POWER USED IN RUNNING HYDRAULIC ELEVATORS.

Q. The question lately came up between some engineers as to the proper method of estimating the power that should be used in running hydraulic elevators. For instance, A claims (the car of the elevator being counterbalanced) that the average load in the car only must be taken into consideration in connection with the height it is lifted, making due allowance for resistance or friction, and to illustrate this he gives the following example: An average of four passengers and the operator each trip lifted 100 feet in a minute, the average weight of the persons being 150 lbs. each, or 750 lbs. × 250 lbs., for friction, etc., which will be

$$\frac{1,000 \text{ lbs.} \times 100'}{33,000}$$

or three-horse power. On the other hand, B claims that the maximum load must be considered every time, and gives as his reason the fact that the water-cylinder must be filled every time the car goes up, let the load be large or small, and reasons this way: Water to fill the cylinder, say 400 gallons, admitted with a pressure due to a 100-foot head of water, which will be,

$$\frac{3,000 \text{ lbs. water} \times 100'}{33,000}$$

or nine horse-power, to which must be added the loss of power and steam due to elevating the water by common pumps, which work irregularly and intermittently. Who is correct?

A. The whole amount of water elevated to the tanks or lowered in the cylinder must be taken into account each trip. B takes the proper view of the matter, but it must be remembered that it is only while the elevator is going up that the water is being used from the tank, and that only for one-half the time can the water be assumed to be used; therefore, presumably 4.5 horse-power is all that is actually used, so far as the elevating of the water is concerned, but probably as much more power is wasted; so B's estimate of the whole power to be accounted for is more nearly correct.

MELTING SNOW IN THE STREETS BY STEAM.

Q. Can snow be melted economically by the use of a steam contrivance such as was tried this winter in the streets of New York?

A. When we find people still anxious to demonstrate that they can melt snow in the street with less cost in time, money, and muscle than it can be shoveled and carted away for, we are urged to review this question with the hope of saving the inventor's money, and perhaps some municipal authorities' expense, though we are safe in saying that should the latter move in this matter with the caution they generally do in similar matters, and try their apparatus before they buy it, they will find that the cost of fuel will always exceed the cost of shoveling and carting, unless, indeed, the city is immensely large and on a plain, without a river or other suitable place to dump snow in.

The heat necessary to warm and melt one pound of snow from 10 degrees above zero to water at 32 degrees above zero Fahrenheit is 164 heat units. The greatest reasonable amount of heat to be obtained and utilized by the burning of one pound of coal is 10,000 heat units, which gives us $\frac{10000}{164}$, or very nearly 61 pounds of snow as the greatest theoretical quantity that can be melted for such a day as January 9, 1886. But in practice some of this snow is turned into vapor, and the heat necessary to convert one pound of it to vapor is 1,168 heat units, so that if we consider only one pound of it is made into vapor, it will reduce the amount of snow melted to less than 54 pounds, and each additional pound of snow converted into vapor will lessen the snow melted per pound of coal an additional seven pounds; so that, presumably, 30 pounds of snow melted per pound of coal is more than actual practice can ever obtain by any rapid method such as by portable machines.

Taking the ordinary snowfall, then, for the day as equivalent to one inch of rain-water (an extraordinarily low estimate) over a mile of street 75 feet between the house-lines, we have 2,062,500 pounds of snow, or 68,750 pounds of coal, or $171.87 worth of coal to a mile of street. But it is not necessary to go so far; it is only necessary to consider that the burning of 34½ tons of coal on a mile of street even in a day is a task in itself not to be considered.

THE ACTION OF ASHES STREET-FILLINGS ON IRON PIPES.

Q. WILL you please inform me what action, if any, coal-ashes have upon iron? Can such ashes be used as "street-filling" without injury to iron, water, or gas service-pipes beneath them? Does the *lye* from said ashes injure the pipes? With what material can said pipes be coated to prevent such injury?

A. Coal-ashes cannot be used for street-filling without injury to iron pipes which are placed in such material. Indeed, moist coal-ashes form about the most destructive filling which could be used. This arises largely from the presence of sulphur compounds in the ashes, and not from any caustic or carbonated alkalies. In some cases the pipes are attacked in spots and slowly eaten through, while in other instances the entire pipe becomes rotten, and converted into a soft graphitic substance.

Pipes may be covered on the outside with pitch, which is some protection, although probably of little permanent benefit. Pipes are sometimes laid in pitch, in addition to the coating of that material. A bed and covering of common soil is also some protection.

ARRANGEMENT OF STEAM-COILS FOR HEATING OIL-STILLS.

Q. I HAVE a number of tanks or stills in which I evaporate a light coal-oil. Within these tanks are steam-coils, to which high-pressure steam is admitted at full boiler pressure. We had difficulty in evaporating—that is, our coils did not work properly—and we were advised to put in new coils, with larger diameter pipes, and make a "gravity return apparatus" of the whole. This we did, and we cannot see that we are much better off than before.

When we first start up a still it takes hours to get it warm. We are forced to close the return-pipe to the boiler and blow our live steam through the coils in great quantities, wasting it and reducing the steam-pressure for other purposes. After blowing in this way for two or three hours at short intervals the thing gets gradually to work, and we have little more trouble until we have to start again.

Can you suggest a remedy or give the cause of such action? We were informed that a gravity apparatus was a panacea for all troubles in steam-work. Why does it not work here?

A. A gravity apparatus is good in its place, but yours is evidently a place for some other system.

You ask if we can suggest a remedy or give the cause of the trouble. We will consider the cause first. In "Steam Heating for Buildings," page 171, the author says, in speaking of steam-kettles: "The connections should be large and the return-water pipe should not be put back into a gravity circulating apparatus, but should be carried away by a good steam-trap."

In speaking of vacuum-pans that cook or boil by steam heat passed through spiral coils, he says: "When a quantity of water is to be raised from ordinary temperatures to boiling, it must be borne in mind that it will take *of steam* at least *one-fifth* of the weight of the water in the pan to raise it (the water) to the boiling point, and that, when steam is first turned into the coils of the pan, the *shrinkage—i. e.*, condensation of the steam—is enormous, and the result will be the filling up of the space within the coil with the condensed steam" (water formed). He explains that this is due to the rapid loss of pressure of steam as it advances through the coil; the loss of pressure being due to the sudden condensation by the steam, which is brought in contact with a mass of cold water. The absence of pressure, then, within the coil allows the "back" water from the boiler or main return-pipes to rush up within the coil, or should this water be held down by a check-valve and be prevented from backing up, the condensed steam will fall on to the check-valve and fill up within the coil, but without sufficient pressure on its surface to force it through the check-valve. He also says: "That while the great difference of temperature between the water in the still and the steam in the coil lasts, the coil can be warmed a comparatively short distance, leaving only the first short part of the coil that is heated to boil the water in the kettle or still."

You evidently experience the trouble he anticipates. At first you cannot heat up. You then "blow through" and get rid of the condensed water that you have not sufficient pressure behind to force into the boiler, but which will run off against atmosphere. This allows the steam to pass into the coil and warm the water until such time as the coil again fills up. Then you blow the water out again, and eventually the oil in the stills warms up and becomes of the same temperature as the steam, after which the condensation of the steam is proportional only to the amount of heat necessary to maintain the evaporation in the stills.

We would advise the use of a pump operated by a pump-governor to return the water to the boiler by mechanical action

CONVERTING A STEAM APPARATUS INTO A HOT-WATER APPARATUS AND BACK AGAIN.

Q. I HAVE a steam-warming apparatus in my house which I constructed myself from information I obtained by reading the *Sanitary Engineer* and anything else I could find on the subject of house-warming.

It works very well in cold weather, but in ordinary weather, in spring and fall, and sometimes in winter, the house is too warm, and I am forced to shut off some of the coils to cool the house. This does not do it properly, as the rooms into which the heaters are closed may then not be warm enough.

Can I change or arrange a steam apparatus so that it may be run either as a steam or hot-water apparatus, and is it advisable?.

What I wish to accomplish is to run the apparatus as a hot-water one in mild weather, assuming I will be able to get sufficient heat from coils at 180° to 150°, or even lower.

Will hot water for the same temperature do as well as steam in pipes?

A. In some cases and with some classes of radiators this may be done.

With vertical or inclined tube-radiators, fastened at one end to a base, this cannot be done. If, on the other hand, the heaters are box-coils or hollow castings, such as "Gold Pin," "Compound Coil," or

"Clogston's" indirect, or heaters of that class, or the "Bundy" hot-water, "Reed Three-Column," or direct heaters of these classes, it may be done, if the fitter understands the principles of hot-water heating.

What the result will be in any particular case will always be doubtful until tried, but if any person desires to alter a steam apparatus already in use he must proceed as follows:

From the highest point in the steam-main run a small pipe, say three-quarters of an inch inside diameter, to a small tank on the top of the house. From any other point in the steam-mains or heaters in which air can lodge or collect run a similar small pipe to the same tank, or to a larger pipe terminating in the tank, care being taken that the air cannot lodge in the pipes, but may free itself by gravitation into the tank. These pipes are all air-vents when thus put in, and will be then automatic, but each should have a valve in it so that it may be closed tightly when the apparatus is to be used for steam.

This tank is best when almost closed from the atmosphere, and should have a capacity of at least one-twentieth of the cubic contents of the whole apparatus—boiler, pipes, and heaters—to allow for the expansion of the water from say 40° to 212°, or thereabouts. When starting the apparatus it must be filled with cold water until it shows in this tank.

If there is a check-valve in the return-pipe, remove it so as to present no obstruction to the flow of water.

The higher the tank can be placed above the heaters the better, as by this means a great pressure is maintained in the boiler, and a wider limit secured before the point of making steam is reached.

The value of the heating-surface will be the same for like temperatures maintained whether steam or hot water is employed. But with hot water and small pipes the temperature will be much less than with steam, as the water rapidly parts with its heat and has to depend on its circulation to the boiler for a renewed supply.

CONDENSATION PER FOOT OF STEAM-MAIN.

Q. CAN you inform a reader of your journal the amount of condensation per square foot of surface that takes place in the pipes of the New York Steam Company, or in steam-pipes properly laid under ground?

A. We are not aware that the New York Steam Company, or any one for them, ever made public the loss of steam due to condensation in their pipes in the streets. In any case the loss of heat will largely depend on the method of protecting the pipes from moisture in the ground, and the material they may be covered with to prevent loss of heat by radiation and conduction.

The Holly Company, of Lockport, N. Y., states in a circular that in 1,600 feet of 3-inch pipe protected in their manner and laid on a descending grade of twenty feet, with the lower end trapped for water and a constant steam-pressure of twenty pounds maintained at each end as near as it was possible to measure it, that during twelve hours the water condensed was eighty-two pounds per hour. This represented 932 heat units per pound weight of steam condensed from steam at twenty pounds pressure to water at the same pressure; thus, $82 \times 932 \div$ square feet of surface of pipe $(1,468) = 52.06$ *heat units* per square foot of surface per hour. This is .179 of a pound of steam at twenty pounds pressure condensed per square foot of actual outside surface of pipe.

The manner of preparing the pipe and the conditions under which it was placed are given as follows: The pipe is wound with asbestos, followed by hair-felting, porous paper, Manilla paper, and finally thin strips of wood laid on lengthwise, and the whole wound by a copper wire and thrust into a wooden log bored to leave an air-space between the pipe and the log. The whole is laid in a trench in the ground, and an earthen drain-pipe placed below it to carry off water from the ground.

With higher pressures the condensation of course will be greater per unit of surface, increasing in a ratio presumably about as the increase of pressure of the steam.

OIL IN BOILERS FROM EXHAUST STEAM.

Q. WE would like your opinion on a question of danger from oil in boilers, and to be understood we will explain as follows : A building is heated by exhaust steam from engine. The condensed water is returned to a tank, then pumped into mud-drum of boiler. Water returned will give the boiler about eight-tenths of its supply; city water (which has some lime in it) is added to make the required amount used. Mineral-oil is used to lubricate engine, cylinder, valves, etc. Suction-pipe from pump to tank is arranged to pump not closer than six inches to the bottom of tank and not to take water within two inches of the surface of water—that is, it is arranged not to pump any sediment or anything that would float in tank. We want to use return-water, as it is thought to be best for boiler, but some persons have told the owner that using water in this way will cause the boiler to burn above fire, the oil forming a bubble or otherwise preventing the water from reaching or absorbing the heat from boiler-plate. For years, after cleaning out boiler, the same oil now used to oil cylinder has been put in boiler, a gallon at each time, to prevent scale. The same city water has been heated in a Stillwell heater (a surface-heater with plates to direct water so it falls through exhaust steam), then pumped to boiler. We admit that animal fat or oil in a boiler will do great harm—that is, enough of it will—but even cylinder-oil with animal matter in it, where water is returned from heating-apparatus, will do no more harm than where a (Stillwell) surface-heater is used. The oil has just as good a chance to get in boiler one way as the other. We have written the details so you would have a better understanding, but the question is : We say using exhaust steam for heating, pumping the water of condensation back into the boiler, where cylinder is lubricated with mineral-oil, etc., (tanks cleaned every two weeks) will not injure boiler, but will be better than fresh water that has lime in it ; but we also admit that it is not the proper thing to pump water of condensation back to boiler without extra care in cleaning tanks where *animal* fat or oil is used to lubricate cylinder, etc. The other party says that the steam containing the oil being passed through heating-apparatus repeatedly changes the nature of the oil and water, and causes it to do harm to boilers, but also says putting the same oil directly into the boiler through manhole, a gallon at a time, will not harm the boiler, but will clean it. (That is the custom here—to put in black oil to prevent scaling). The question then narrows to this ;

Will it harm a boiler to pump the water of condensation (using exhaust steam in heating-apparatus) back to boiler repeatedly, using mineral oil to lubricate cylinder, etc.?

A. Of the danger of carrying most, if not all, qualities of oils into a boiler along with the feed-water from any source there is no question. The publishers of the *Locomotive*, who are probably in as good a position as any persons in the country to obtain knowledge on this subject, are afraid of the results that may follow the use of oil. They do not object to the use of *pure mineral* oil—meaning, presumably, crude petroleum—and say : " Crude petroleum is one thing, but that *black oil*, which may mean almost anything, is very likely to be something different." To be safe, nothing but the crude petroleum, that can be purchased for less than ten cents per gallon, should be used as a boiler purge.

About the danger of animal oils, and probably also of the vegetable oils there seems to be not the least question, but many do not hesitate to use large quantities of these cheap manufactured mineral oils for boiler purging, not knowing, or not considering, that many of these cheap oils are a little better than a residuum of some other manufactured products, and containing many, if not all, of the heavy constituents of crude petroleum and other substances that may be added to give a "body" or increase the lubricating power of the oil.

More on this question may be found on page 33, where an illustration of the *misuse* of oil in boilers is given.

On the other hand, exhaust steam from cylinders of engines or pumps can be and is put into the general heating-apparatus, and the water thus formed pumped into the boilers without injury. It must be remembered, though, that while this is possible with a properly arranged apparatus, it is very easy to arrange one that may work injury.

The oil must be separated from the steam if it is carried over, and this separation should be done before the exhaust steam enters the heating-pipes, otherwise the pipes and radiators must become the recipients of a considerable portion of the oil or grease. The inside of the pipes form a kind of condensing-surface for the heavy oils that are not held in suspension, and only very light oils, as a general thing, ever reach the receiving-tank. If oil is found in the tank, your method is the only one we know of to prevent its getting into the boiler.

Frequent and conscientious inspection of the tank and boiler is all that can be then done by the engineer, who, if he finds the grease accumulates in the tank in such quantities that some of it finds its way into the boiler, must then consider means of preventing it, either by separation or by wasting the steam to atmosphere, or condensing it in a separate part of the heating-pipes and allowing this portion of the condensed water to go to waste.

Where the water contains lime it is important that all the condensed water should be recovered, and for this reason: If you find by actual examination that no grease is carried into the boiler, or, at least, none can be found there (as the lime may have something to do with neutralizing a little oil), that you do right to do so.

On page 171 will be found a description of an apparatus used to separate oil from exhaust steam, and which may be used by any one, as it is not patented.

INDEX.

AIR-BINDING in return-pipes and the remedy, 83, 85.
Air-binding of box-coils and the remedy, 59.
Air, capacity of to contain watery vapor, 120.
Air compressed in tanks applied to the automatic raising of water, 116, 118.
Air, arrangements to remove from steam-apparatus which is used as hot-water apparatus, 220.
Air, quantity of required for heating and ventilating rooms of known dimensions, 81.
Air, sensible temperature of, as affected by the presence of watery vapor, 122.
Air-spaces in walls, 121.
Air, theoretical and actual velocites of in flues, 97.
American Steam Company's system of street-mains, 203, 204, 205.
Animal oils in boilers dangerous, 223.
Asbestos cardboard for gaskets, 23.
Ashes in coal, percentage of, 136.
Ashes in street fillings, their action on iron pipes, 217.
Aspirating-shafts in the Ogdensburg, N. Y., opera-house, 196.
Austria and Germany, methods of heating houses in, 198 to 203.
Automatic apparatus for raising water in buildings, 116 to 119.
Automatic pump-governor, 137.

BAKER, SMITH & Co.'s sectional radiator, 140, 141.
Baldwin, W. J., recommends removable boiler-lugs, 29.
Bending pipes and cutting nipples, 106 to 115.
Bending pipes in the shop, 111.
Blowing off and filling a boiler in summer, 17.
Boiler, horizontal, for house-heating, re-setting of, 153.
Boiler plant arranged for heating several adjacent buildings, 71, cf. 81, cf. 130.
Boilers, 17 to 44.
Boilers, applying filters and other means to the purification of water supplied to, 131.

Boilers, blowing off and filling during summer when the boiler is not to be used, 17.
Boilers, calculating the quantity of water in, 155.
Boilers, carbonic-acid gas accumulating in and causing danger of suffocation, 25.
Boilers, domes supposed to have caused charring of wood, 40.
Boilers for hot-water and steam apparatus, amount of heating or fire surface in, 51.
Boilers for hot-water heating in England, 158, 159.
Boilers, feeding of, size of pumps, and forcing-pipes necessary for, 26.
Boiler foundations, 154.
Boilers, isolating-valves for, 29 to 33.
Boilers, lugs for arranged to be removable, 28.
Boilers, magazine muzzles of. (See magazines.)
Boilers of the Manhattan and Merchants' Bank Building, 172.
Boilers of the Mutual Life Insurance Building, 178, 179.
Boilers of the Tribune Building, 182 to 187.
Boilers, oil in, dangers of, 33, 222.
Boilers, pipes from. (See pipes.)
Boilers, position of try-cocks on, 155.
Boilers, proper place on to place test-gauges, 18.
Boilers, range, life of, 42.
Boilers, rivets for, iron and steel, 35, 36.
Boilers, rivets. (See also rivets.)
Boilers, safety-valves for. (See safety-valves.)
Boilers, setting of in the Tribune Building, 182 to 187.
Boilers, several connected together, accident from careless treatment of an expansion-joint and valve-connections, 39.
Boilers, size of estimated for given radiator-surface, 52.
Boilers, water in, what will be condition of when used repeatedly in a gravity-apparatus, 38.
Boilers, water in, expanding by heat, 21.
Boilers, water-line of, how low should it be, 82.

Boilers, water-line of, how to determine, 155.
Boilers, water-tube, in France, 206.
Boiling as a means of purifying water, 134.
Boston Water-Works, ram for testing fittings at, 134.
Box-coils, air-binding of and the remedy, 59.
Brass pipe, rate of expansion of, 213.
Breakage of steam-mains in New York streets, 205, 206.
Buildings, several adjacent, heating with a single boiler plant, 71, cf. 81, cf. 130.

CALCULATING quantity of water in a boiler, 155.
Carbonic acid gas in boilers causing death of an inspector, 25.
Cast-iron safe for steam-radiators, 138.
Cast-iron surface compared with pipe-surface for heating, 62.
Centennial Exhibition, experiments on ashes in coal, 136.
Centrifugal fan used in France, 210.
Charring of wood said to be caused by boiler-dome or steam-pipes, 40 to 42.
Church, pilasters in utilized for flues 188.
Church, ventilation of, 98.
Churches, warming of by coils in each pew, 65, 67.
Circuits established improperly between steam-risers and return-pipes, 214.
Cisterns for hot-water heating, arrangement of in large buildings, 43.
Cisterns, of copper, of galvanized-iron, rusting of, etc., 42.
Cisterns. (See also tanks.)
Close nipples of large size, how to cut, 106.
Close nipples, removing from couplings after the thread is cut, 110.
Coal and radiator surface, relation between, 154.
Coal consumption in Hauber's patent stoves, 203.
Coal, amount of required for heating certain buildings or rooms, 45, cf. 54.
Coal fires, advantages of feeding new coal above or below them, 201.
Coal, percentage of ashes in, 136.
Coal, per square foot of grate in economical consumption, 154.
Coal-tar coating for pipes, 130.
Coils, box, air-binding of and the remedy, 59, 63.
Coils connected with engine-exhaust should not be of too small diameter, 102.

Coils compared with pipe-surface, 62.
Coils for hot-water heating in France, 208, 209.
Coils for superheating steam, length of required, 101.
Coils, remarks on the proper setting and connection of, 63.
Coils, amount required for rooms of known dimensions, 88.
Coils, ¾-inch, connected with engine-exhaust, 102.
Coils for oil-stills, 217.
Cold air ducts, sizes computed, 89.
Condensation of vapor on walls, 120.
Condensation of steam per square foot of pipe-surface in the pipes of the New York Steam Company, 221.
Condenser for steam-exhaust pipes to prevent the fall of spray, 144 to 146.
Condensed water, method of estimating the heat due to, 152.
Connected boilers, accident caused by careless treatment of expansion-joint in the connection of one of the boiler-valves, 39.
Connecting steam and return risers at their tops, reasons for, 213, 214.
Cocks, try, on boilers, 155.
Copenhagen Fine Arts Exhibition Building, heating of, 192 to 194.
Cost of steam for melting snow on the streets, 216.
Cost of steam for warming a given room or space, 54, cf. 45.
Cover of cloth for regulating a steam-radiator, 56.
Crooked threads, how to cut them on close nipples, 108.
Cushing's, Frank A., automatic pump-governor, 137.
Cutting crooked threads on close nipples, 108.
Cutting 4-inch close nipples, 106.
Cutting nipples and bending pipes, 106 to 115.
Cutting nipples of large size, 106, 112.
Cutting threads of various sizes with a solid die, 113.

DAKOTA Apartment-House, cast-iron safes in for radiators, 138.
Dampness on walls, and its prevention, 120.
Detroit, Mich., First Baptist Church, heating of, 66.
Diameters of standard pipe, 74.
Dies, solid, adapted to cutting threads of different sizes, 113.
Differential ram for testing fittings, 134.

INDEX.

Direct radiation, cost as compared with indirect, 45, 46.
Direct radiation for church-warming, 65, 67.
Dome of a boiler supposed to have charred wood, 40.
Domes, purposes and utility of discussed, 20.
Domes, size of sheets and character of steel for making, 19.
Double glazing, economical results of, 49.
Ducts for warm air in the auditorium of the Ogdensburg, N. Y., Opera-House, 195, 198.
Dye-houses, arrangement of flues and pipe surface to remove vapor from, 92.

ELEVATOR pump-connections with a steam-trap so arranged as to waste steam and cause back-pressure, 128.
Elevators, hydraulic, computing the power necessary to raise water for, 215.
Elmira, N. Y., State Reformatory, ventilation and heating of, 148.
Emery's, C. E., isolating-valve, 32, 33.
England, low-pressure hot-water system of heating in, 156.
Exhaust-condensers for preventing the fall of spray, 144 to 146:
Exhaust steam and live steam used in the same heating job, 99.
Exhaust steam carrying oil into boilers, effects of, 222.
Exhaust steam for heating the Fine Arts Exhibition Building in Copenhagen, 192 to 194.
Exhaust steam for heating the Manhattan and Merchants' Bank Building, 165 to 168.
Exhaust steam for heating, its economy, 99.
Exhaust steam from a given engine, amount of surface required to condense it, 103.
Exhaust steam not to be turned into too small a coil, 102.
Exhaust steam, tank for separating grease from, 170.
Expansion and apparent increase of bulk of water in boilers due to heat, 21.
Expansion-joints designed to prevent telescoping, 153.
Expansion-joints on mains in New York streets, 203, 204, 205, 206.
Expansion-joint in the connection between two boilers causes an accident, 39.
Expansion of pipes of various metals, 75.

Expansion of brass pipe and of iron pipe, 213.
Expansion of steam-pipes, apparently less than theory requires, 75.
Experiments on the percentage of ashes in coal, 136.
Explosion of steam-table, 103; means to prevent, 104.
Extended surface, definition of 61.
Extended surface heaters or radiators used in France, 207.

FAN for church ventilation, size of, 98.
Fan, centrifugal, used in France, 210.
Fan, helicoidal, used in France, 210, 211.
Fan, hydro-ventilator, used in France, 211.
Feed-pipes, different arrangements of suggested where the amount of water fed was insufficient, 26, 27.
Feed-water carrying oil into boilers, effects of, 223.
Feeders for boilers, applying filters to, 131.
Feeding boilers, difficulties arising from insufficient pumping power, great length, and insufficient size of pipes, 26.
Filters for boiler-feeders, 131.
Fires, are they likely to be caused by steam-pipes, 40, 42.
Fitting and piping, 70 to 85.
Fitting, just good enough to work, 70, 71.
Fittings, differential ram for testing, 134.
Flues devised in a church by use of the pilasters, 188.
Flues heated by Bunsen gas-burners, 190.
Flues, size of computed, 87, 88, 89, 97.
Flues. (See also ducts and aspirating-shafts.)
Foundations for boilers, 154.
Fractional valves for graduating radiator-surface, 142.
Fractional valves in the Manhattan and Merchants' Bank Building, 164, 169, 170.
France, heating and ventilation of buildings in, 206 to 211.
Fuel, amount of required to warm buildings of certain capacities, by hot-water heating, 47.
Fuel, amount of required to warm buildings of certain capacities, on direct and indirect radiation, 45, 47, cf. 54.

GAS-BURNERS, Bunsen, used for heating ventilating-flues, 190.

228   INDEX.

Gas-pipes, testing for leaks in, 132.
Gas-pipes, making joints on tight, 133.
Gaskets of asbestos cardboard preferable to those of India rubber for safety-valve connections, 23.
Gauges for testing boilers, where to be placed, 18.
Geneste, Herscher et Cie's pamphlet on the heating and ventilation of buildings in France, 206.
Glass, relation of radiating surface to, 49.
Globe-valves, proper and improper positions of on radiators, 57.
Gold's, E. E., method of graduating radiator surface, 143.
Gold's pin surface, comparative value of, 62.
Governor, automatic, for pumps, 137.
Graduating radiator surface, 139 to 144.
Graduating-valves in the Manhattan and Merchants' Bank Building, 164, 169, 170.
Graduating valves. (See also fractional valves.)
Grate surface to give economical combustion of fuel, 154.
Gravity apparatus, how to apply a steam-trap to, 212.
Gravity-return heating-apparatus boilers, condition of water in, 38.
Gravity-return heating-apparatus, used for several adjacent buildings, 71, 81.
Grease-separating tank in the Manhattan and Merchants' Bank Building, 170; in the Mutual Life Insurance Co.'s Building, 180.

HAUBER'S patent stoves in Germany, 200.
Heat due to condensation of water, method of estimating in order to test work of steam-apparatus, 152.
Heat in low-pressure and high-pressure steam available for heating purposes, comparison between, 100.
Heat given out by steam-apparatus, its relation to heat due to condensation, and method of estimating, 152.
Heating and ventilation of buildings in France and other countries, 206 to 211.
Heating and ventilation of the Ogdensburg, N. Y., Opera-House, 195 to 198.
Heating and ventilation of the State Reformatory at Elmira, N. Y., 148.
Heating and ventilation of the West Presbyterian Church in New York City, 187 to 192.
Heating and ventilation of the "Umbria," 93.

Heating-apparatus in the Manhattan and Merchants' Bank Building, 161.
Heating-apparatus in the Fine Arts Exhibition Building in Copenhagen, 192 to 194.
Heating-apparatus, steam, in the Kalamazoo Insane Asylum, 146.
Heating-apparatus on the gravity-return principle, condition of the water in boilers of, 38.
Heating by exhaust-steam, economy of, 99.
Heating by low-pressure hot-water apparatus in England, 156; in the United States, 160.
Heating by hot water, radiators for, 58.
Heating certain buildings and the amount of fuel required, 45, cf. 54.
Heating churches by direct radiation, 65, 67.
Heating houses in Germany and Austria, 198 to 203.
Heating on the one-pipe system, 79.
Heating several buildings from same boiler plant, 71, cf. 81, 130.
Heating-surface, how much will a steam-pipe of given size supply, 52.
Heating-surfaces or fire-surfaces, amount of required in hot-water-apparatus and steam-apparatus boilers, 51.
Heating-surfaces, pipe and cast-iron, relative value of, 62.
Heating-surfaces proportioned to air-space, 45.
Heating-surfaces proportioned to a given room, 51, 53.
Heating-surfaces required to condense steam from a given engine-exhaust, 102.
Heating-surfaces, value of, 45 to 55.
Heating water for Hotel Warren, 126.
Heating water for large institutions, 125.
Heating water in large tanks, how to do it, 124.
Helicoidal fan used in France, 210.
Holly Company's estimates of condensation in steam-mains, 221.
Holly Company's steam-pipes, method of laying, 221.
Hood's theory of hot-water circulation, 157.
Hopkinson, J., & Co.'s isolating-valve, 29, 30.
Horizontal boiler, resetting of, 153.
Hot-air flues, size of computed for rooms of given dimensions, 87, 88, 89.
Hot-water apparatus and steam-apparatus arranged to be interconvertible, 219.
Hot-water circulation, Hood's theory of, 157.

INDEX.

Hot-water heating, amount of surface required, 47.
Hot-water heating-apparatus in France, 208, 209.
Hot-water heating, low pressure, in England, 156; in the United States, 160.
Hot-water heating, radiators for, 58.
Hot-water heating, stove arranged for, 60.
Hotel job of steam-fitting, with mains too small, no reliefs, and other defects, 70.
Hotel Warren, in Boston, method of heating water for, 126.
House-heating systems in Germany and Austria, 198 to 203.
Hydraulic elevators, computing the power required to raise the necessary quantity of water, 215.
Hydro-ventilator fan used in France, 211.

INSPECTORS of boilers endangered by the presence of carbonic-acid gas, 25.
Iron pipe injured by ashes in street fillings, 217.
Iron pipe, rate of expansion of, 213.
Isolating-valves where several boilers are used: the English valve, 29 30; the American valve, 31, 33.

JOINTS, expansion, designed to prevent "telescoping," 153.
Joint, expansion, in the connection between two boilers, causes an accident, 39.
Joints on gas-pipes, making them tight, 133.

KAL, term used by New York Steam Company, meaning of, 55.
Kalamazoo Insane Asylum, steam-heating apparatus in, 146.

LEAKS in gas-pipes, testing for, 132.
Lochiel Hotel, steam-table at explodes, 103.
Locomotive, The, on the use of oil in boilers, 33.
Low-pressure hot-water system in England, 156.
Lugs for boilers arranged to be removable, 28.

MAGAZINE MUZZLES on house-heating boilers, how to prevent their burning off, 21.
Manhattan and Merchants' Bank Building, steam-heating apparatus in, 161; boilers in, 172.
Melting snow in the streets by steam, estimate of the fuel required, 216.

Meigs, General M. C., on the economical results of double glazing of windows, 49.
Mills system patents, what are they? 83.
Miscellaneous, 124 to 211.
Miscellaneous questions, 212 to 224.
Moisture, effect of on sensible temperature, 122.
Moisture on walls, effect on sensible temperature, 120 to 123.
Moisture on walls, causes and prevention, 120.
Mutual Life Building, steam-heating apparatus in, 171; boilers in, 178, 179.
Muzzles of magazines, how to prevent their burning off, 21.

NEW YORK STEAM COMPANY furnishes power and heat to the Mutual Life Insurance Building, 178, 179, 180.
New York Steam Company's system of street-mains, 204.
New York Steam Company's mains, condensation in, 221.
New York streets, system of steam-pipes in, 203 to 206.
Nipples, close, of large size, how to cut, 106, 113.
Nipples, close, cutting crooked threads on, 108.
Nipple, close, removing from a coupling after the thread is cut, 110.
Nipple-cutting and pipe-bending, 106 to 115.
Noise in steam-pipes, cause of, 78.

OGDENSBURG, N. Y., opera-house, warming and ventilation of, 195 to 198.
Oil entering boilers from exhaust steam, is it harmful? 222.
Oil carried into boiler with feed-water, effects of, 223.
Oil in boilers, dangers of use of, 33.
Oil-stills, arrangement of steam-coil for, 217.
Oil, separating from exhaust-steam, 223, cf. 171.
Oils, animal and cheap mineral, dangerous in boilers, 223.
One-pipe system of steam-heating, 79.
Overhead piping and the advantages claimed for it, 76.

PASCAL IRON-WORKS, rules for determining sizes of flues, 89.
Patents of the Mills system, what are they? 83.
Percentage of ashes in coal, 136.
Pilasters of a church utilized for flues, 188.

Pipe-bending and nipple-cutting, 106 to 115.
Pipe-bending in the shop, 111.
Pipe, iron, injured by ashes in street fillings, 217.
Pipe-guards on the "Umbria," 95, 96.
Pipe of standard sizes, true diameters and weights of, 74.
Pipe, steam. (See steam-pipe.)
Pipe surface compared with cast-iron surface for heating, 62.
Pipe surface compared with coils, 62.
Pipe-threads. (See threads.)
Pipes, coating with coal-tar, 130.
Pipes, expansion of apparently less than theory requires, 75.
Pipes, brass and iron, expansion of, 213.
Pipes from boilers, proper method of passing through walls, 24.
Pipes of various metals, rates of expansion of, 75.
Pipes. (See also steam-pipes.)
Piping and fitting, 70 to 85.
Piping in a church heated by direct radiation, 68.
Piping in the Manhattan and Merchants' Bank Building, 162, 165, 166; in the Mutual Life Co.'s Building, 182.
Piping, overhead, the advantages claimed, 76.
Piping where several adjacent buildings are heated by same boiler plant, 71, 82.
Plane surface, or plain surface, difference of meaning of, 61.
Plates of boilers when of steel should be riveted with steel rivets, 35, 36.
Plates of different thicknesses, and corresponding proportions of rivets, 37.
Plenum system of ventilation in the Kalamazoo Insane Asylum, 146.
Power required to raise water for hydraulic elevators, 215.
Pressure-regulating valve used in France, 206, 207.
Pressure regulation in the Mutual Life Company's Building, 180.
Preventing fall of spray from steam exhaust-pipes, 144 to 146.
Prison at Elmira, ventilation and heating of, 148.
Public institutions, how to heat water for, 125.
Pump-engine of elevator connecting with steam-trap so as to waste steam and cause back-pressure, 128.
Pump-governor, automatic, 137.
Pumping hot water and steam, 72, 74.
Pumps for feeding a battery of boilers, size of required, 26, 28.

Pumps or traps in a job where several buildings are to be heated from one boiler plant, 73; relative economy of, 73.
Purifying water by boiling, 134.
Purifying water for boilers, 131.

RADIATION, direct and indirect, compared with reference to fuel required, 45, 46.
Radiating-surface. (See radiator-surface.)
Radiator, safe for, of cast-iron, 138.
Radiator, sectional, 139, 140, 141.
Radiator-surface, amount of required for hot-water heating of certain buildings, 47.
Radiator-surface, amount of required for steam or hot-water heating of certain buildings, 49.
Radiator-surface, amount of proportional to glass, 49.
Radiator-surface, methods of graduating, 139 to 144.
Radiator-surface, relation of to coal-consumption, 154.
Radiator-surface required for a given room, 51, 52, cf. 88.
Radiator-surface proportional to size of boiler, 52.
Radiators and heaters, 56 to 69.
Radiators and long coils, relative efficiency of, 53.
Radiators, direct-indirect, in a prison, 149.
Radiators for hot-water heating, what forms will answer, 58.
Radiators, what kinds can be used for hot-water heating, 219.
Radiators, regulating, woman's method of, 56.
Radiators, regulating. (See graduating radiator-surface.)
Radiators used in France, 207.
Radiators, valves on, improperly placed, 57.
Raising water automatically, 116 to 119.
Ram on the differential principle for testing fittings, 134.
Reck's, A. B., system for heating the Fine Arts Exhibition Building in Copenhagen, 192 to 194.
Registers, sizes of for rooms of given dimensions, 86, 97.
Regulating steam pressures in the Mutual Life Insurance Co.'s Building, 180.
Regulating-valves used in France, 206, 207; in Mutual Life Building, 180.
Resetting house-boiler, 153.
Return-pipes, how to run when obstacles are met, and how to obviate air-binding, 85.

INDEX. 231

Return-pipes, air-binding in and the remedy, 83, 85.
Return-pipes and main steam-pipes connected at the top, 213, 214.
Riser connections, proper place of valves on, 77.
Risers, steam and return, connected at the top, 213, 214.
Rivets of iron in steel boiler-plates, objections to, 35.
Rivets of steel in steel boiler-plates, good results from use of, 36.
Rivets, proportions of, with reference to thickness of plates. 37.
Rooms, amount of radiator-surface required for, 51.

SAFE of cast-iron for steam-radiators, 138.
Safety-valves rendered inoperative by careless placing of gaskets, 22.
Safety-valves, importance of proper connection with boilers, 22.
Sanitary Engineer, The, office heated by hot-water apparatus, 161.
Sectional radiators, 139, 140, 141.
Setting boilers in the Tribune Building, 182 to 187.
Setting house-boiler, 153.
Sky-light in the Manhattan and Merchants' Bank Building, arrangement of steam-pipes about, 167.
Slotting magazine muzzles to prevent their burning off, 21.
Snow in the streets, melting by steam, cost of fuel for, 216.
Solid dies adapted to the cutting of threads of different sizes, 113, 114.
Specification for boilers in the Manhattan and Merchants' Bank Building, 172 to 176.
Spray from steam-exhaust pipes, condensers to prevent fall of, 144 to 146.
Standard pipe, true diameters and weights of, 74.
Steam, 99 to 105.
Steam-apparatus so arranged as to be convertible into a hot-water apparatus and back again, 219.
Steam-coils for heating oil-stills, 217.
Steam, condensation of in pipes of New York Steam Company estimated, 221.
Steam, exhaust. (See also exhaust steam.)
Steam exhaust carrying oil into boilers, effects of, 222.
Steam, exhaust, economy of using for heating, 99.
Steam for melting snow, computing cost of, 216.
Steam-fitting in a hotel, with mains too small, no reliefs, and other defects, 70.

Steam for power and heating in the Mutual Life Insurance Company's Building derived from street-mains of the New York Steam Company, 178, 179, 180.
Steam-heating apparatus in the Fine Arts Exhibition Building in Copenhagen, 192 to 194.
Steam-heating apparatus in the Kalamazoo Insane Asylum, 146.
Steam-heating apparatus in the Manhattan and Merchants' Bank Building, 161.
Steam-heating apparatus in the Mutual Life Building, 177.
Steam-heating apparatus, method of ascertaining heat given off by, through determining heat due to condensation of water, 152.
Steam-heating on the one-pipe system, 79.
Steam, low-pressure and high-pressure, available heat in for heating, 100.
Steam-pipe, how much heating surface will a $\frac{3}{4}''$ pipe supply, 52.
Steam-pipes and return-pipes connected at the tops, 213, 214.
Steam-pipes in New York streets, method of laying, expansion-joints, etc., 203 to 206; breakage, 205, 206.
Steam-pipes, expansion of apparently less than theory requires, 75.
Steam-pipes, condensation in, estimated, 221.
Steam-pipes, condensers to prevent the fall of spray from, 144 to 146.
Steam-pipes, noise in and its cause, 78.
Steam-pipes of Holly Company, method of laying, 221.
Steam-pipes supposed to cause charring of wood and fires, subject discussed, 40 to 42.
Steam-radiator. (See radiator.)
Steam riser connections, proper position of valves on, 77.
Steam, superheating by coils, 101.
Steam-table, explosion of, 103.
Steam-tables arranged to diminish liability of explosion, 103, 104.
Steam-trap connected with pumping-engine so as to waste steam, 128.
Steam-trap, how to be applied to gravity apparatus, 212.
Steam-trap, proper connection of with engine cylinder, 129.
Steam-trap, how the pipe conveying live steam to it must be connected with the boiler, 213.
Steel for domes and boilers, 19.
Stills, oil, steam-coils for, 217.

Stove arranged for hot-water heating, 60.
Stoves for heating houses in Germany and Austria, 198, 199.
Stoves, Hauber's patent, in Germany, 200.
Street fillings of ashes injurious to iron pipes, 217.
Street-mains as a source of power and heat in the Mutual Life Insurance Co.'s Building, 178, 179, 180.
Streets of New York, steam-pipe systems in the, 203 to 206.
Streets, snow in, cost of melting by steam, 216.
Stumpf's, G., apparatus for raising water automatically by compressed air, 116.
Suffocation of workmen in boilers, 25.
Superheating steam by coils, 101.
Supervising inspectors for steam-vessels, their rules for placing test-guages, 18.
Surfaces, extended, definition of, 61.
Surfaces for heating, how much will a steam-pipe of given size supply, 52.
Surfaces for heating (fire surfaces) in hot-water and steam-apparatus boilers, 51.
Surfaces for heating proportioned to air-space, 45.
Surfaces for heating proportioned to a given room, 51, 53.
Surfaces for heating, value of, 45 to 55.
Surfaces, pipe and cast-iron, relative value of, 62.
Surfaces, plane or plain, meaning of terms, 61.
Surfaces, radiator. (See radiator surfaces.)
Switch-valves in the West Presbyterian Church, in New York City, 192.

TABLE, steam, explosion of, 103.
Tailoring and steam-fitting, "just good enough" work in both, 71.
Tanks for boilers and for hot-water heating, 42.
Tanks, capacity of one of given dimensions, 127.
Tank for separating grease from exhaust steam in the Manhattan and Merchants' Bank Building, 170; in the Mutual Life Insurance Co.'s Building, 180.
Tanks, heating water in, 124, 126, 128.
Tanks, size of supply to fill in a given time, 127.
Tar, coal, as a coating for pipes, 130.
Temperature, sensible, affected by moisture in the air, 122.
Test-gauges, where to be placed on boilers, 18.

Testing fittings by differential ram, 134.
Testing gas-pipes for leaks, 132.
"Thermus" on connecting steam and return risers at the top, 213, 214.
"Thermus" on cutting crooked threads on close nipples, 108.
"Thermus" on cutting large close nipples, 106.
"Thermus" on removing large close nipples from couplings after the thread is cut, 110.
Threads, crooked, cutting them on close nipples, 108.
Threads, cutting on large close nipples, 106.
Threads of various sizes, cutting with the some solid die, 113.
Trap, steam, connected with elevator pumping-engine so as to waste steam, 128.
Trap, steam, how to be applied to the gravity apparatus, 212.
Trap, steam, proper way to connect with engine-cylinder, 129.
Traps or pumps in a job where several buildings are to be heated from one boiler plant, 73; relative economy of, 73.
Tribune Building, steam-exhaust pipe condensers for preventing the fall of spray, 145; setting of the boilers in, 182 to 187.
Trowbridge, Prof. W. P., on heated flues or fans for securing currents in flues, 93.
Try-cocks, proper positions of on boilers, 155.
Tudor's, Frederick, fractional-valve for graduating radiator-surface, 142.

"UMBRIA," Cunard steamer, ventilation and heating of, 93.

VACUUM-PANS, arrangement of steam-coils for use with, 218.
Valves for isolating any of several boilers, importance of and patterns of used in England and America, 29 to 33.
Valves for regulating steam-pressure as used in France, 206, 207.
Valves, fractional, for graduating radiator-surface, 142, 143.
Valves, fractional, in the Manhattan and Merchants' Bank Building, 164, 169, 170.
Valves, globe. (See globe-valves.)
Valves improperly placed on radiators, causing the bases to fill with water, 57.
Valves, switch. (See switch-valves.)

INDEX.

Valves, proper and improper places for on steam-riser connections, 77.
Vapor condensing on walls and its prevention, 120.
Vapor in the air affects sensible temperature, 122.
Vapor, removing from dye-houses by means of heated flues, 92.
Varnishes for making joints on gas-pipes tight, 133.
Ventilating-flues, constructing by utilizing pilasters of a church, 188.
Ventilating-flues heated by Bunsen gas-burners, 190.
Ventilation, 86 to 98.
Ventilation and heating of buildings in France and other countries, 206 to 211.
Ventilation and heating of the Ogdensburg, N. Y., Opera House, 195 to 198.
Ventilation and heating of the West Presbyterian Church, in New York City, 187 to 192.
Ventilation and heating of New York State Reformatory, at Elmira, 148.
Ventilation by heated flues or by fans, 93.
Ventilation-flues, computing sizes of, 89.
Ventilation of a church, 98.
Ventilation of Cunard steamer "Umbria," 93.
Ventilation, plenum, in the Kalamazoo Insane Asylum, 146.
Ventilators, window, for schools, etc., 90, 91.

WALLS, moisture collecting on and its prevention, 120.
Walls, proper method of passing pipes through, 24.
Walworth Manufacturing Co.'s method of graduating radiator-surfaces, 143.
Warming and ventilation of buildings in France and other countries, 206 to 211.
Warming and ventilation of the Ogdensburg, N. Y., Opera House, 195 to 198.

Warming and ventilation of the West Presbyterian Church, in New York City, 187 to 192.
Warming. (See also heating.)
Water, boiling of as a means of purifying, 134.
Water condensed in steam-apparatus indicative of heat given off, method of conducting tests, 152.
Water, condition of in boilers of gravity apparatus, 38.
Water, expansion of and apparent increase of bulk when a boiler is fired up, 21.
Water for boilers, methods of purifying, 131.
Water for hydraulic elevators, power required to raise, 215.
Water-hammer, cause of, 78.
Water, heating, in large buildings, 43.
Water, heating, in large institutions, 125.
Water, heating, in large tanks, 124.
Water, heating, in large tanks, quantity of steam or hot water required to do it, 127.
Water in a boiler, calculating the quantity of, 155.
Water-line, how low should it be to insure satisfactory working, 82.
Water-line, how to determine for different boilers, 155.
Water, method of heating for the Hotel Warren, 126.
Water, method of heating on the "Umbria," 96.
Water-motor for driving fan in France, 211.
Water, raising automatically in buildings, 116 to 119.
Water-pipe, coating with coal-tar, 130.
Water-tube boilers in France, 206.
Weights of standard pipe, 74.
West Presbyterian Church in New York City, heating and ventilation of, 187 to 192.
Window-ventilators, 90, 91.
Windows, double glazing of, economical results of, 49.
Woman's method of regulating a steam-radiator, 56.

THE SANITARY ENGINEER, devoted to ENGINEERING, ARCHITECTURE, CONSTRUCTION, AND SANITATION, is a weekly journal for the Architect, Engineer, Mechanic, and Municipal Officer.

OPINIONS OF THE PRESS.

" THE SANITARY ENGINEER is one of the very best publications of its kind in this or any other country, and may be regarded as the representative paper devoted to architecture and engineering."—*Boston Herald.*

" Stands at the head of all publications of its class."—*Chicago Tribune* (March 30, 1886).

" Has very ably filled the position it has made for itself."—*New York Commercial Advertiser* (February 19, 1886).

" The recognized authority on all questions pertaining to architecture, engineering, construction, and sanitation."—*Indianapolis Journal.*

" The special illustrations are very beautiful, and the articles are written by the most competent and practical scientific men. There are carpenters and builders in this city who assert that they would not be without this periodical even if it contained nothing more than advertisements, as that department contains information of all the latest and most desirable inventions and improvements in building and sanitary supplies and apparatus."—*The Bridgeport Standard* (April 16, 1886).

" Standard authority on matters pertaining to its specialty. It will be found an instructive addition to the library of the practical man as well as the engineer."—*American Machinist.*

" THE SANITARY ENGINEER brings something useful and interesting every week, but a collection of the issue for six months together impresses one anew with the value of its work and the intelligence with which it is edited."—*Hartford Courant.*

" A large and flourishing weekly journal, covering the whole field of Sanitary Science and recognized as a leading authority upon the subject."—*The Nation.*

" THE SANITARY ENGINEER shows an excellent appreciation of what may be done in the field of Sanitary Engineering, and a practical ability for doing it."—*N. Y. Tribune.*

" The leading journal of the kind published in the United States."—*Troy Times.*

Sold by Newsdealers, 10 cents a copy.

Subscription, post-paid, United States and Canada, $4.00 per annum. Great Britain, 20 shillings. Foreign Countries in the Postal Union, $5.00.

Published every Thursday.

No 140 WILLIAM STREET, NEW YORK.
92 and 93 FLEET ST., LONDON.

www.ingramcontent.com/pod-product-compliance
Lightning Source LLC
Chambersburg PA
CBHW021820230426
43669CB00008B/811